The Genera of Fungi
Sporulating in Pure Culture

by

J.A. VON ARX

second, fully revised edition
with 134 figures
illustrating more than 300 fungi

1974 · J. CRAMER
In der A.R. Gantner Verlag Kommanditgesellschaft
FL-9490 VADUZ

1st edition 31.VIII.1970
(Reprinted 1973)
2nd edition 5.VIII:1974

© 1974 A. R. Gantner Verlag KG., FL - 9490 Vaduz
Printed in Germany
by Strauß & Cramer GmbH, D-6901 Leutershausen
ISBN 3 7682 0693 9

Contents

INTRODUCTION

Most keys for the identification of fungi are designed to be used with material that has been freshly collected from nature or with herbarium specimens. Therefore many of the characteristics in these keys refer to the mode of growth on the natural substrate, in many cases this is a host plant. On the other hand, keys for the identification of fungi in culture exist only for small or large groups.

Some years ago the author drew up keys for the identification of fungi in pure culture and these appeared to be adequate in the practical work of students in mycology. To make them available to a larger public they were later published in German (von Arx, 1967). It became apparent that after the genus had been found with the help of these keys the identification of the species was often difficult because the pertinent literature could not be easily found.

In the present English version these difficulties are alleviated by giving the references to the descriptions of the accepted genera. Furthermore, the type species is given and usually the relevant data on the synonyms. Should a monograph or other study of the given genus be known, this is mentioned in the references and the number of species described is indicated.

First, a discussion of the natural relationships between the fungi is given to clarify the system used here. In a later chapter the most important spore types known in the fungi are arranged in a simple system.

The following have been omitted from the keys and the lists of genera: the Acrasiales and the Labyrinthulales of the Myxomycota, the Tuberales, the Erysiphales, most of the 'Discomycetes' and the Laboulbeniales of the ascomycetes, and all the orders of the basidiomycetes.

Most of the representatives of the Acrasiales that have been investigated in pure culture belong to the genera *Dictyostelium* Brefeld and *Polysphondylium* Brefeld which are difficult to distinguish (see Bonner, 1967; Raper, 1935, 1951). Only a few Myxomycetes are known in culture on agar media; these were cultivated on water agar with oatmeal in the presence of bacteria or yeasts. Well known for instance are the plasmodial cultures of *Physarum polycephalum* Schw. of *Badhamia utricularis* (Bull.) Berk. One of the few species that form sporangia in culture is *Physarum didermoides* (Pers.) Rostaf. Cultures of Labyrinthulales are as yet hardly avaiable.

The cultivation of species belonging to the Tuberales seems to be difficult or completely unsuccessful. The formation of tuberous ascomata has not yet been achieved on sterilized media. The fungi belonging to the Erysiphales are exclusively biotrophic parasites of higher plants and only grow on the appropriate host. The same also applies for many other ascomycetes with perithecia or apothecia. The Laboulbeniales are restricted to insects.

Most of the basidiomycetes often have a characteristic mycelium in pure culture but fructifications with basidia and basidiospores have been obtained from only a few, usually coprophilous, species. These are mostly members of the Aphyllophorales and Agaricales, belonging for example to the genera *Coprinus* (Pers. ex Fr.) S. F. Gray, *Clitocybe* (Fr.) Kummer, *Pholiota* (Fr.) Kummer, *Agaricus* L. ex Fr., *Schizophyllum* Fr. and *Exobasidiellum* Donk. Many other basidiomycetes form typical conidia in culture, usually arthrospores (called oidia), as well as frequent hyphal swellings or chlamydospores. The mycelial mat is often well developed, lanate and white or light coloured, and may be dark in old cultures. In many Aphyllophorales especially, various types of hyphae occur, e.g. either with or without clamp connections, also skeletal hyphae with thick walls, very fine binding hyphae or coloured gloeohyphae. Nobles (1948, 1965) has produced keys based on cultural characteristics with which one can identify many wood-attacking basidiomycetes.

The Uredinales (rust fungi) are exclusively biotrophic plant parasites and either cannot or only with great difficulty be cultivated in pure culture. It is impossible to identify the species without knowing the respective host plant.

Only those genera whose members grow and reproduce in pure culture are given in the keys. Mainly those genera represented by the cultures present in the Centraalbureau voor Schimmelcultures (CBS) have been included. In the meantime many other fungi will have been obtained in pure culture, thus the keys and lists are not claimed to be complete. In fact they are only a beginning and later there will be many additions.

The reader will miss a division in families. However in many groups families are unsufficiently described. The keys do not pretend to give a natural classification at all; they rather are prepared for practical use.

Without the unstinting help and advice of all the members of the CBS the keys and lists would have been much less complete. The author owes especial thanks to Dr. W. Gams, Mr. H. A. van der Aa, Miss W. Ch. Slooff and Mr. D. Yarrow. The last mentioned also corrected the English text.

The author is greatly indepted to Dr. E. Müller (Zürich), Dr. G. L. Hennebert (Louvain), Dr. W. B. Kendrick (Waterloo, Ontario), Dr. C. Booth (Kew) and Mr. G. H. Boerema (Wageningen) for helpfull suggestions and advice during the writing of this publication.

April 1970   Centraalbureau voor Schimmelcultures
                                                Baarn, The Netherlands

connects them with the Chytridiomycetes, but there are no true intermediates.

The Mycota, the true fungi, can be characterized as fungi which produce hyphae or, as in the yeasts, only blastospores or arthrospores. The cell wall usually contains chitin (polymerized glucosaminacetate) and never cellulose. Chitin is absent only in a few yeasts (*Schizosaccharomyces*). Motile spores are unknown in the Mycota, this in contrast to the chytrids. True gametes are also absent, but conidia may function secondarily as gametes.

The following classes are distinguished:

a. **Zygomycetes** with a coenocytic, multinuclear mycelium. The copulation takes place between gametangia, forming a thick-walled resting spore, the zygospore. Fruiting bodies in general are unknown.

Two orders have to be mentioned, the Entomophthorales and the Mucorales. The Entomophthorales contain parasites on insects and other animals. The spores are produced exogenously on conidiophores and often are shot away when ripe.

The Mucorales in general are saprophytic, the asexual spores are produced in sporangia or sporangioles, or by more highly evolved forms on conidiophores and are not shot away.

b. **Endomycetes**. In this class are included all yeasts and yeast-like fungi and some plant parasitic fungi hitherto placed in the Ascomycetes, the Basidiomycetes and the Deuteromycetes. In the Endomycetes, fruiting bodies or hymenia are absent. The mycelium is usually also absent or reduced. These fungi however produce blastospores, sprout-cells or arthrospores, fission cells or sometimes a pseudomycelium. In pure culture they all grow in a yeast-like manner.

In the Endomycetes we have to consider a number of orders, e.g. the Endomycetales with the sporogenic or ascosporogenic yeasts (hitherto placed in the Ascomycetes) and the Torulopsidales for the asporogenic or imperfect yeasts. The Taphrinales and the Ustilaginales are exclusively plant-parasitic fungi. All species of these two orders which have been obtained in pure culture grow in a yeast-like manner, producing blastospores. These states are often described as imperfect yeasts. As plant parasites these fungi develop within the living tissue of the host without killing it. The Ustilaginales are characterized by thick-walled resting spores, which after germination produce haploid sprout-cells, the sporidia. The germ tube, called promycelium, is often compared with the basidia of the Basidiomycetes. The Ustilaginales are related to otomyces, a genus of the Taphrinales. *Protomyces* also has resting sp... But after germination, the haploid sprouting cells are produced endog... in a sac-like projection. The *Protomyces* species can be considered as ... in which the sporidia are borne in an ascus-like cell of the resting... cultures of the two groups cannot be distinguished.

Well known and well characterized are the ...ree remaining classes of the Eumycota:

c. **Ascomycetes** possess a septate, hapl... mycelium and asci which are produced in fruiting bodies, the asco... may be apothecia, cleistothecia or peri-

5

thecia. On the basis of the structure of the ascus, we can distinguish between unitunicate and bitunicate Ascomycetes. A large number of orders have been established.

d. **Basidiomycetes** are characterized by a septate, frequently dicaryotic mycelium with or without clamp connections. Basidia are produced in basidiocarps or in hymenia on the dicaryotic hyphae.

On the basis of the basidial structure, one distinguishes between the Holobasidiomycetidae with one-celled basidia and the Phragmobasidiomycetidae with septate basidia. The latter contain especially the rust fungi and their relatives. Some taxonomists tend to exclude the Uredinales from the Basidiomycetes and to put them in a new class. Although this might be more correct, it does not appear to be necessary, because the Uredinales have no closely related forms within the other fungi.

e. **Deuteromycetes**, the imperfect fungi, have a septate mycelium and no known sexual forms. The asexual spores may be produced on the hyphae, on conidiophores or in fruiting bodies, but not in sporangia.

Since most mycologists for practical reasons prefer to classify all fungi and fungus-like organisms in a single, polyphyletic kingdom, the following arrangement appears to be convenient as given by von Arx (1967):

kingdom: Mycota

phylum: Myxomycota
classes: Myxomycetes
Acrasiomycetes
Plasmodiophoromycetes
Labyrinthulomycetes

phylum: Oomycota
classes: Oomycetes
Hyphochytridiomycetes

phylum: Chytridiomycota
class: Chytridiomycetes

phylum: Eu-Mycota
classes: Zygomycetes
Endomycetes
Ascomycetes
Basidiomycetes
Deuteromycetes

THE FUNGUS SPORE

The most typical feature of the fungi is undoubtedly the mycelium, consisting of hyphae or filaments. These can be coenocytic or septate, hyaline or dark, simple or united into strands. Besides the vegetative mycelium the fruitbodies are often composed of hyphae or hyphal elements. In Basidiomycetes especially, several types of hyphae can be distinguished. Apart from the yeasts the hyphae therefore can be an important character for the identification of a fungus.

Likewise, the spores are a typical feature of the fungi. These reproductive structures serve for multiplication, dispersal, and survival. The morphological character and especially the manner of development of the spores represent the most important feature for the identification of a fungus. Some fungi cannot be identified in pure culture because they lose their sporulating capacity.

Fungus spores can be motile (planospores, swarm cells, zoospores) or non-motile (aplanospores). **Planospores** move in water mostly with the help of flagella. Fungi with planospores are mostly watermoulds or occur under humid conditions as soil saprophytes or plant parasites. Planospores are known especially in Myxomycetes, Chytridiomycetes and Oomycetes, but do not occur in the Mycota (Eu-Mycota).

Aplanospores for dispersal are dependent on air currents, streaming water or animals, especially insects. Many spores are violently discharged or show other adaptions for dispersal (Ingold, 1953).

In the ontogeny of the spores two main types can be distinguished: those borne endogenously in cells or those borne exogenously either on or from cells.

Endogenous spores are borne mostly in specialized cells, the **sporangia**. Motile spores are generally borne in sporangia called zoosporangia and motile sex cells (gametes) also are borne in sporangial cells, the **gametangia**.

Sporangia forming non-motile sporangiospores are typical of the Mucorales. Small sporangia containing only one or a few spores are often called **sporangiola**.

The **asci** of the ascomycetes are specialized sporangia, in which after meiosis a constant number of aplanospores is usually produced, e.g. 8 or 4. In typical ascomycetes the asci arise on dicaryotic hyphae in fruiting bodies called ascomata.

Endogenous spores can also be formed in hyphal cells or (in sporogenic yeasts) in sprout cells. In hyphal cells these spores are often arranged in one row. In yeast cells the number of endospores is limited, occasionally to one. Often such endospores have a thick wall and they function as resting spores. Spores may also be formed endogenously in the meristematic zone of a specialized sporogenous cell in basipetal succession and then they may be freed through an opening or fissure of the sporogenous cell (e.g. phialospores and arthrospores).

Exogenous spores can be borne by sprouting or by the formation of a cell wall followed by fission. Sprouting cells are called **blastospores**, fission cells are **arthrospores** (**blastoconidia** and **thalloconidia** sensu Kendrick, 1971).

There are intermediates between blastospores with a small base and arthrospores with a cross wall without any constriction between the mother and daughter cell. A spore may arise as a blown-out end of a cell by budding with a broad base and moderate constriction, and the final separation of two cells takes place by the formation of a cross-wall. Such a spore may be considered as a blastospore with a broad base or as an arthrospore with a constriction.

Blastospores with a narrow and with a broad base and arthrospores are known in yeasts (Lodder and Kreger-van Rij, 1952). Species of the genus *Saccharomyces*

7

form blastospores with a small base. In this genus the sporulation is multipolar all over the surface. In *Schizosaccharomyces* the multiplication takes place by fission. Blastospores with a broad base followed by the formation of a septum are typical in genera such as *Nadsonia* or *Saccharomycodes*. In this case the formation of daughter cells takes place only at the ends of the mother cells (bipolar) and can repeat itself by a meristematic growth of the apical region. A number of daughter cells can be produced in basipetal succession as blown-out ends or by proliferation through the scars of previous daughter cells. In some cases a collarette-like sheath may surround the scar. In the Deuteromycetes (Fungi Imperfecti) blastospores of different kinds can be distinguished. Blastospores with a narrow base may be borne simultaneously all over the surface of a sporogenous cell. Or the spores are borne in succession direct on the hypha or on short, denticulate outgrowths. The conidiophore may successively form new growing points to one side of the previous conidium. The conidiophore, therefore, either increases in length or becomes apically thickened and covered with denticles. Blastospores with a narrow base may be separated from the sporogenous cell by a total constriction as the small lumen of the canal between sporogenous cell and conidium fills up with cell-wall substances.

Blastospores with a broader base also may be borne acropetally on conidiophores which increase by growing sympodially at one side, becoming geniculate and covered with conidial scars (Radulasporae, Sympodulosporae).

Blastospores with a narrow or braod base can produce a new blastospore at the apex, thus forming acropetal chains of conidia, which may be simple or branched. In chains of blastospores the points of attachement are usually broad allowing the transport of nutrients. In fully grown chains, thickened cross walls with or without disjunctors often develop between the conidia (compare *Cladosporium* or *Monilia*). The conidia in this type mostly are comparable with the bipolar multiplication in some yeasts.

Blastospores with a narrow or a broad base can also be borne in basipetal succession on a specialized sporogenous cell. Such a cell develops one or more open ends from which the conidia arise. If the opening is narrow and the conidia have a small, rounded base, the sporogenous cell is called a phialide and the conidia are **phialospores**. The phialide may be attenuated towards the opening in the tip, which may or may not possess a collarette (e.g. *Phialophora*). If the meristematic zone is broad, often each new conidium develops by successive proliferation through the scars of previous conidia. In such cases the sporogenous cell may elongate and become annellate, and it is called an **annellophore**. Blastospores with a broad base may also be borne on a sporogenous cell with an apical collarette-like sheath. In this case the sporogenous cell does not elongate (compare the conidial states of *Monascus* or *Eremascus*).

Porospores are blastospores borne singly through pores of thick-walled conidiophores. Porospores mostly are dark and thick-walled and may occur in acropetal chains. There are intermediate forms to genera such as *Cladosporium* with chains of blastospores with thickened scars of attachment.

Basidiospores are borne in the same manner as blastospores mostly with a narrow base. They arise after meiosis on spicula-like processes (pedicels) of a specialized cell, the basidium. The pedicels are called sterigmata; they mostly develop in a constant number of 4 (rarely 2, 6 or 8), depending on the number of haploid daughter-nuclei. Basidiospores are mostly asymmetrical, and have a hilum (a lateral mark at the point of attachment) and are forcibly discharged when mature (ballistospores). A similar mechanism of forced discharge occurs in *Sporobolomyces, Tilletiopsis* and other imperfect, mostly yeast-like fungi where the ballistospores arise on vegetative cells or on hyphae. Spores borne on sterigmata develop always singly and never in succession.

Arthrospores develop by a simultaneous or basipetal septation (fragmentation) of sporogenous hyphae or conidiophores and are arranged in simple or branched chains. They may also arise endogenously by the formation of cylindrical spores within sporogenous hyphae. They become free by histolysis or breaking up of the outer wall of the sporogenous hyphae. Such spores were occasionally described as ascospores in *Endomyces (Geotrichum)* and related genera. In *Thielaviopsis* and related genera cylindrical conidia arise endogenously in basipetal succession by fragmentation of the protoplasm in a meristematic zone and are liberated in chains through an apical opening.

Aleuriospores or **chlamydospores** usually are thick-walled; they arise as a terminal, lateral or intercalary swelling of a conidiophore or a hypha and are separated from the mother cell by one or two septa. Aleuriospores mostly are persistent and function as resting spores. In other cases aleuriospores may arise as a terminal swelling of a tapering conidiophore. They are freed easily by the formation of a separating cell or a disjunctor.

Blastospores with a broad base and borne in basipetal succession are often characterized as aleuriospores. In the imperfect fungi, especially in the Moniliales, Hughes (1953), Tubaki (1958), Subramanian (1962) and Barron (1968) distinguished 8 or more sections by the manner in which conidia are formed and by the structure of the conidia-bearing apparatus. This classification is preferable to the classification given by Saccardo (1886) and based on conidium morphology. The classifications given by the 4 authors mentioned above differ in some respects. The preferable arrangement seems to be that given by Barron, because he distinguishes clearly between aleuriospores and blastospores with a broad base (called annellosporae). But in general, the distinction between blastospores with a narrow and blastospores with a broad base, is only applied in yeast taxonomy. Therefore in the taxonomy of Moniliales closely related genera such as e.g. *Spilocaea* and *Fusicladium* are usually classified in different sections although both genera have blastospores with a broad base.

So much for the mechanisms of spore formation. The spores themselves can be divided into Xero- and Gloiosporae. **Xerosporae** are formed in dry masses and then are often arranged in rows or chains or represent dusty masses (Wakefield & Bisby, 1941; Mason, 1937), which are dispersed by air currents. **Gloiosporae** form wet, slimy or viscous droplets or masses and are dispersed by water. especially by rain, or by insects. In one and the same species, the

9

spores can be wet or dry and taxonomically this character is little of value.

Spores can be 1-celled (amerosporae) or septate. In the latter case we distinguish between didymosporae (2-celled), phragmosporae (with 2 or more transverse septa), dictyosporae (with transverse and longitudinal septa), scolecosporae (filiform or vermiform), staurosporae (radiate or stellate), helicosporae (spirally coiled). Spores can be colourless, hyaline (Hyalosporae) or pigmentated mostly brown or blackish (Phaeosporae). These characters were used by Saccardo (1886) in his arrangement of Hyphomycetes and other fungi. But in one and the same species, spores can be 1-celled or septate, hyaline or coloured. Therefore the system of Saccardo is not even of practical use; colour and cell number of spores is considered by modern taxonomists as being of minor value. Nevertheless these characters and the shape and size of the spores have to be used frequently in distinguishing genera or species.

KEY TO THE ORDERS OF FUNGI

1. Motile spores (zoospores) or myxamoebae are present (absent only in higher Oomycetes with cell walls containing cellulose), copulation between motile gametes or between an egg cell and either a motile gamete or an antheridium 2

1. Motile spores or myxamoebae are absent, the cell walls never contain cellulose 10

2. Thallus naked, a myxamoeba or a plasmodium 3

2. Thallus with cell walls 6

3. Plasmodia and resting spores develop in the cells of a host plant (obligate endoparasites), zoospores biflagellate, flagella of unequal lenght . . . .
. . . . **Plasmodiophorales** p. 15

3. Plasmodia and resting spores do not develop in a host cell 4

4. Mass of resting spores powdery, thallus is a multinucleate, free-living plasmodium, zoospores present or absent, uni- or biflagellate . . . .
. . . . **Myxomycetales**

4. Mass of resting spores slimy, thallus is an uninucleate myxamoeba 5

5. Myxamoebae form a pseudoplasmodium, mass of spores slimy, zoospores absent . . . . **Acrasiales**

5. Myxamoebae form a network of slimy filaments, zoospores mostly present (aquatic, mostly marine) . . . . **Labyrinthulales**

6. Zoospores posteriorly uniflagellate, flagellum of the whip-lash type 7

6. Zoospores anteriorly either uni- or mostly biflagellate, one flagellum is of the tinsel type, cell walls contain cellulose, copulation between an egg cell and an antheridium 9

7. Thallus small, with or without rhizoids, a nucleate mycelium is absent . .
. . . . **Chytridiales** p. 16

7. Thallus mostly a coenocytic branched mycelium (but some *Blastocladiella* species have a spherical, quite large thallus without mycelium) . . .8

8. Copulation takes place between motile gametes, resting sporangia present, mostly ellipsoidal . . . . **Blastocladiales** p. 20

8. Copulation takes place between a non-motile egg cell and a motile gamete, oospores thick-walled, mostly spherical . **Monoblepharidales** p. 22

9. Zoospores anteriorly or laterally biflagellate . . . . **Oomycetales** p. 25

9. Zoospores small, anteriorly uniflagellate . . . . **Hyphochytridiales** p. 24

10. Sporocarps, hymenia and usually also a true mycelium are absent, thalli are sprout or fission cells with nuclei, cultures yeast-like 13

10. Mycelium coenocytic or septate, sporocarps, hymenia or sporangia mostly present 11

22. Ascomata mostly are cleistothecia and develop on a superficial or erumpent mycelium (obligate parasites on higher plants, not known in pure culture)...
.... Erysiphales and Meliolales

22. Ascomata are perithecia, apothecia or cleistothecia, in the latter case not parasitic on higher plants 23

23. Ascomata are perithecia or cleistothecia.............. Sphaeriales p. 104

23. Ascomata are discoid apothecia, or the hymenium is exposed while the ascospores are ripening...................................... 24

24. Ascomata with usually a dark covering layer opening by a slit or by irregular lobes .... Phacidiales p. 82

24. Ascomata without a dark covering layer, usually superficial or becoming so, hymenium exposed from an early stage .... Helotiales p. 82

25. Ascomata spherical or nearly so, or dimidiate, ostiolate or with an apical pore .... Pseudosphaeriales p. 132

25. Ascomata not ostiolate, opening by slits or by dehiscence 26

26. Asci spherical or nearly so, scattered singly throughout the ascomata ...
.... Myriangiales

26. Asci spherical, clavate or elongate, arranged parallel in one layer........
.... Dothiorales p. 132

27. Basidia septate... 28

27. Basidia not septate 30

28. Septation of basidia vertical or cruciform, sterigmata terminal.........
.... Tremellales

28. Basidia cylindrical, transversely septate, sterigmata lateral 29

29. Basidia borne on fruit bodies .... Auriculariales

29. Basidia borne on teliospores (obligate parasites on vascular plants)......
.... Uredinales

30. Basidia slender, terminal with two horn-like sterigmata .. Dacrymycetales

30. Basidia not so 31

31. Hymenium gymnocarpous, hemiangiocarpous or in a few cases angiocarpous, then basidiocarp fleshy 32

31. Hymenium or gleba angiocarpous, spore mass becoming gelatinous, powdery or cartilaginous, sterigmata often reduced, basidiospores are not ballistospores... 33

32. Trama of the basidiocarp without swollen hyphal cells, mostly firm, not soft or putrescent, hymenium smooth, on teeth, in tubes or rarely on gills .... Aphyllophorales

32. Trama of basidiocarp with swollen cells or cell walls becoming slimy, soft or putrescent, hymenium mostly in tubes or on gills ....
.... Agaricales

3 sub-orders have to be distinguished:

Trama of the basidiocarp heteromerous, including sphaerocysts or/and latex cells, spore ornamentation amyloid. . . . . .Russulineae

Trama of the basidiocarp homomerous, trama of the hymenophore bilateral, hyphae thin, with walls becoming slimy . . . . Boletineae

Trama of the basidiocarp homomerous, cells of the hyphae partly swollen, not becoming slimy . . . . Agaricineae

PLASMODIOPHORALES

All species belonging to this order are obligate parasites on higher plants or on filamentous fungi and do not grow on nutritional media in pure culture. Species of genera such as *Plasmodiophora* Woronin and *Sorosphaera* Schröter develop especially in roots of higher plants; e.g. *Plasmodiophora brassicae* Woronin is the cause of club root of *Brassica* and other crucifers (Karling, 1942).

Species of the following genera are parasitic in Saprolegniaceae or in other saprophytic moulds and have to be listed here:

Resting spores form a mass .... *Woronina* (2)
Resting spores are grouped predominantly in groups of eight .. *Octomyxa* (1)

List of genera

1. **Octomyxa** Couch & al. - J. Elisha Mitchell scient. Soc. *55*: 400 (1939)
   O. achlyae Couch & al. (Fig. 1b)
   second species: *O. brevilegniae* Pendergrass
   References: Sparrow, 1960.

2. **Woronina** Cornu - Ann. Sci. nat., Bot. Sér. 5, *15*: 176 (1872)
   W. polycystis Cornu (Fig. 1a)
   4 species, parasitic in *Saprolegnia, Achlya, Pythium* or green algae
   References: Sparrow, 1960.

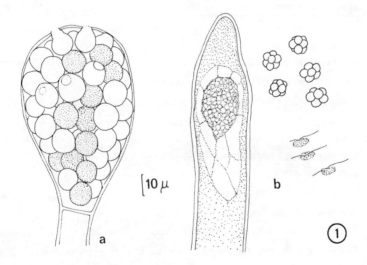

Fig. 1. a, *Woronina polycystis*, swollen hyphal tip of *Achlya* spec. with sporangia; b, *Octomyxa achlyae*, swollen hyphal tip of *Achlya* spec. with plasmodium, groups of resting spores and zoospores (both from Sparrow, 1960).

CHYTRIDIALES

Only a few species, mostly isolated from soil or from plant debris in water, are known to develop in pure culture. Species parasitic on higher plants, belonging to genera such as *Synchytrium* de Bary & Woronin or *Physoderma* Wallr., are biotrophic and develop only on a suitable host plant. Species belonging to other genera are parasitic on algae or on other fungi, especially on Saprolegniales. The most important genera with species growing saprophytically on nutrition media or parasitically in hyphae of other fungi can be distinguished by the following key:

1. Thallus is a single, uniaxial, unbranched filament with a basal holdfast and an attenuated apex 12

1. Thallus is not an uniaxial filament 2

2. Sporangia operculate, opening by a lid 11

2. Sporangia not operculate, opening by pores, tubes or a fissure 3

3. Sporangia endobiotic in the cells of the host 4

3. Sporangia epibiotic, superficial 7

4. Sporangium completely filling the host cell and assuming its shape . . . .
. . . . *Rozella* (10)

4. Sporangium not completely filling the host cell 5

5. Sporangium with more than one discharge tube . . . . *Pleotrachelus* (6)

5. Sporangium with one pore or tube 6

6. Sporangia formed in clusters . . . . *Pringsheimiella* (7)

6. Sporangia scattered, not in clusters . . . . *Olpidium* (4)

7. Thallus monocentric, developing into a sporangium 8

7. Thallus polycentric, composed of a system of rhizoids with a variable number of sporangia . . . . *Cladochytrium* (2)

8. Sporangia with rhizoids arising from several places on the surface 9

8. Sporangia with rhizoids arising at the base or from an apophysis 10

9. Sporangia large, often with a basal cell, pigmented gametangia mostly present . . . . cf. *Blastocladiella*

9. Sporangia small or irregular in size, often yellow or reddish, rhizoids long, gametangia absent . . . . *Rhizophlyctis* (8)

10. Sporangia with a basal apophysis (a swelling) . . . . *Phlyctochytrium* (5)

10. Sporangia without an apophysis . . . . *Rhizophydium* (9)

11. Thallus monocentric, developing to a sporangium . . . . *Chytridium* (1)

11. Thallus polycentric, composed of a system of rhizoids with many sporangia
. . . . *Nowakowskiella* (3)

12. Thallus fusiform, proliferating . . . . *Harpochytrium* (11)

12. Thallus filamentous, long, often becoming septate . . . *Oedogoniomyces* (12)

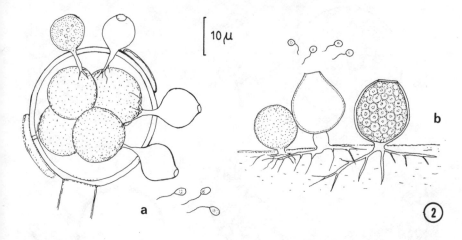

Fig. 2. a, *Rhizophydium carpophilum* (Zopf) Fischer, sporangia on an oogonium of *Achlya* spec. and zoospores (from Coker, 1923); b, *Chytridium olla*, sporangia from a pure culture and zoospores (orig.).

List of genera

(General reference: Sparrow, 1960)

1. **Chytridium** Braun - Betrachtungen über die Erscheinung der Verjüngung in der Natur p. 198 (1851)
   Ch. olla Braun (Fig. 2b)
   about 25 species
   References: Sparrow and Barr, 1955.

2. **Cladochytrium** Nowakowski in Cohn - Beitr. Biol. Pflanzen 2: 92 (1876)
   C. tenue Nowakowski
   9 species
   References: Remy, 1948; Sparrow, 1960.

3. **Nowakowskiella** Schroet. in Engl. & Prantl - Nat. Pflanzenfam. 1, 1: 82 (1892)
   N. elegans (Nowak.) Schroet.
   9 species
   References: Karling, 1944; Gaertner, 1954.

4. **Olpidium** (Braun) Rabenh. in Schroet. - Krypt. Fl. Schles. 1: 180 (1886)
   O. endogenum (Braun) Schroet.
   = *Olpidiella* Lagerh. - J. Bot. Paris 2: 438 (1888)
   = *Endolpidium* de Wildeman - Ann. Soc. Belg. Microb., Mém. 18: 153 (1894)
   E. hormisciae de Wildeman
   about 30 species, e.g. *Olpidium allomycetis* Karling in *Allomyces* spec.
   References: Sparrow, 1960; Sahtiyanci, 1962.

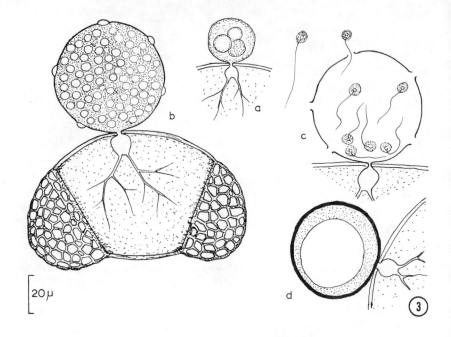

Fig. 3. *Phlyctochytrium spectabile* Uebelmesser, a, young thallus; b, apophysate sporangium on a coniferous pollen; c, sporangium discharging zoospores; d, resting spore (from Uebelmesser, 1956).

5. **Phlyctochytrium** Schroet. in Engl. & Prantl - Nat. Pflanzenfam. 1, *1*: 78 (1892)
   P. hydrodictyi (Braun) Schroet.
   about 30 species, e.g. *P. lippsii* Lohman, parasitic on *Ascobolus, Sordaria* and other ascomycetes; *P. palustre* Gaertner and other species isolated from soil (Fig. 3)
   References: Lohman, 1942; Gaertner, 1954; Übelmesser, 1956; Barr, 1969.

6. **Pleotrachelus** Zopf - Nova Acta Acad. Leop.-Carol. 47: 173 (1884)
   P. fulgens Zopf, parasitic on *Pilobolus* spec.
   about 10 species, e.g. *P. brassicae* (Woron.) Sahtiyanci = *Olpidium brassicae* (Woron.) Dang. on roots of *Brassica* and other plants.
   References: see *Olpidium.*

7. **Pringsheimiella** Couch - J. Elisha Mitchell scient. Soc. *55*: 409 (1939)
   P. dioica Couch, parasitic on *Achlya* spec.
   1 species.

8. **Rhizophlyctis** A. Fischer in Rabenh. Krypt. Fl. 1, *4*: 119 (1892)
   R. mastigotrichis (Nowak.) A. Fischer
   = *Karlingia* Johanson - Am. J. Bot. *31*: 397 (1944)
      K. rosea (de Bary & Woronin) Johanson = *Rhizophlyctis rosea* (de Bary & Woronin) A. Fischer
   about 10 species
   References: Remy, 1948; Gaertner, 1954.

18

9. **Rhizophydium** Schenk - Verh. phys.-med. Ges. Würzburg *8*: 245 (1858)

    R. globosum (Braun) Rabenh. (neotype)

= *Rhizophyton* Zopf - Nova Acta Acad. Leop.-Carol. *52*: 343 (1888)

    R. gibbosum Zopf = *Rhizophydium gibbosum* (Zopf) A. Fischer

about 70 species (Fig. 2a).

References: Barr, 1969, 1973.

10. **Rozella** Cornu - Ann. Sci. nat., Bot., Sér. 5, *15*: 148 (1872)

    R. monoblepharidis-polymorphae Cornu

= *Rozia* Cornu - Bull. Soc. bot. Fr. *19*: 71 (1872) [non Rozea Bescherelle]

= *Pleolpidium* A. Fischer in Rabenh. Krypt. Fl. 1, 4: 43 (1892)

about 20 species, mostly parasitic on water moulds.

11. **Harpochytrium** Lagerh. - Hedwigia *29*: 142 (1890)

    H. hyalothecae Lagerh.

3 or 4 species

References: Emerson and Whisler, 1968.

12. **Oedogoniomyces** Kobayashi & Ôkubo - Bull. nat. Sci. Mus. Tokyo, n.s. *1*: 59 (1954)

    O. lymanaeae Kobayashi & Ôkubo

1 species

References: Emerson and Whisler, 1968.

BLASTOCLADIALES

Only species belonging to the family Blastocladiaceae grow in pure culture.

1. Thallus consisting of a single reproductive body (a sporangium or gametangium), mostly stalked .... *Blastocladiella* (3)

1. Thallus bearing more than one sporangium or gametangium............... 2

2. Thallus composed of creeping hyphae, gametangia present ..... *Allomyces* (1)

2. Thallus composed of branched lobes, often erect, gametangia absent ....
.... *Blastocladia* (2)

List of genera

1. **Allomyces** Butler - Ann. Bot., Lond. *25*: 1027 (1911)
   A. *arbuscula* Butler (Fig. 4)
   = *Septocladia* Coker & Grant - J. Elisha Mitchell scient. Soc. *37*: 180 (1922)
   S. *dichotoma* Coker & Grant = A. *arbuscula*
   about 10 species
   References: Sparrow, 1960; Emerson, 1941.

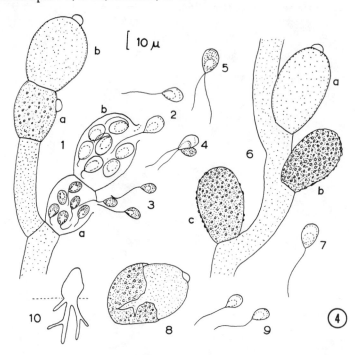

Fig. 4. *Allomyces arbuscula*, 1, part of the gametophyte with (a) ♂ and (b) ♀ gametes; 2, ♂ gametes; 3, ♀ gametes; 4, copulation of gametes; 5, young zygote; 6, part of a sporophyte with (a) zoosporangium and (b, c) resting sporangia; 7, zoospore; 8, germinating resting sporangium, forming haploid zoospores (9); 10, young thallus (orig. from von Arx, 1967).

2. **Blastocladia** Reinsch - Jb. wiss. Bot. *11*: 298 (1878)
   B. pringsheimii Reinsch
   about 15 species
   References: Sparrow, 1960; Emerson and Cantino, 1948.

3. **Blastocladiella** Matthews - J. Elisha Mitchell scient. Soc. *53*: 194 (1937)
   B. simplex Matthews
   = *Clavochytridium* Couch & Cox in Cox - J. Elisha Mitchell scient. Soc. *55*: 389 (1939)
     C. stomophilum Couch & Cox = *B. stomophila* (Couch & Cox) Couch & Whiffen
   = *Rhopalomyces* Harder & Sörgel - Nachr. Ges. Wiss. Göttingen, math.-phys. Kl. *3*: 123 (1938)
                                                                    [non Corda 1839]
     R. variabilis Harder & Sörgel = *B. variabilis* (Harder & Sörgel) Harder & Sörgel
   = *Sphaerocladia* Stüben - Planta *30*: 364 (1939)
     S. variabilis Stüben = *Blastocladiella stuebenii* Couch & Whiffen
   about 10 species
   References: Sparrow, 1960.

MONOBLEPHARIDALES

List of genera

1. **Monoblepharis** Cornu - Bull. Soc. bot. Fr. *18*: 59 (1871)
   M. sphaerica Cornu
   = *Diblepharis* Lagerh. - Bih. K. svenska Ventensk. Akad. Handl. 3, *8*: 39 (1900)
      D. fasciculata (Thaxter) Lagerh. = *Monoblepharis fasciculata* Thaxter
   = *Monoblepharidopsis* Laibach - Jb. wiss. Bot. *66*: (1927)
      M. ovigera (Lagerh.) Laibach = *Monoblepharis ovigera* Lagerh.
   about 10 species (Fig. 5 a-c)
   References: Sparrow, 1960.

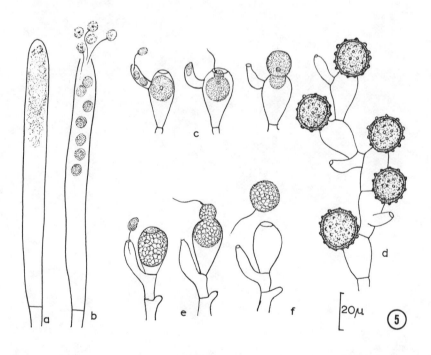

Fig. 5. a, b, *Monoblepharis macrandra* (Lagerh.) Woron., sporangia; c, *Monoblepharis polymorpha* Cornu, gametangia; e, f, *Monoblepharella taylori*, gametangia, planozygote (from Sparrow, 1960).

2. **Monoblepharella** Sparrow - Allan Hancock Pacific Expeditions, Publ. Univ. S. Calif. *3* (6): 103 (1940)

  M. taylori Sparrow (Fig. 5e, f)

  5 species

  References: Sparrow, 1960.

3. **Gonapodya** Fischer in Rabenh. Krypt. Fl. 1,*4*: 382 (1892)

  G. prolifera (Cornu) Fischer

  2 species

  References: Sparrow, 1960.

HYPHOCHYTRIDIALES

Thallus monocentric, simple .... *Rhizidiomyces* (1)
Thallus polycentric, consisting of hyphae .... *Hyphochytrium* (2)

List of genera

1. **Rhizidiomyces** Zopf - Nova Acta Acad. Leop.-Carol. *47*: 188 (1884)
   R. apophysatus Zopf (Fig. 6)
   5 species
   References: Nabel, 1939; Sparrow, 1960.

2. **Hyphochytrium** Zopf - Nova Acta Acad. Leop.-Carol. *47*: 187 (1884)
   H. infestans Zopf
   3 species
   References: Sparrow, 1960; Barr, 1970.

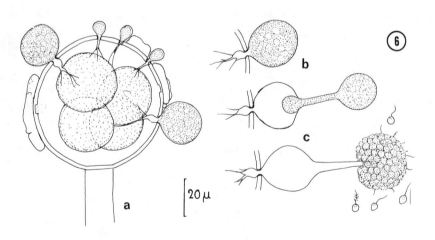

Fig. 6. *Rhizidiomyces apophysatus*, a, superficial thalli with rhizoids in an oogonium of *Achlya* spec.; b, full sized sporangium; c, ripe sporangium germinating with a vesicle and forming zoospores (from Coker, 1923).

OOMYCETALES

26

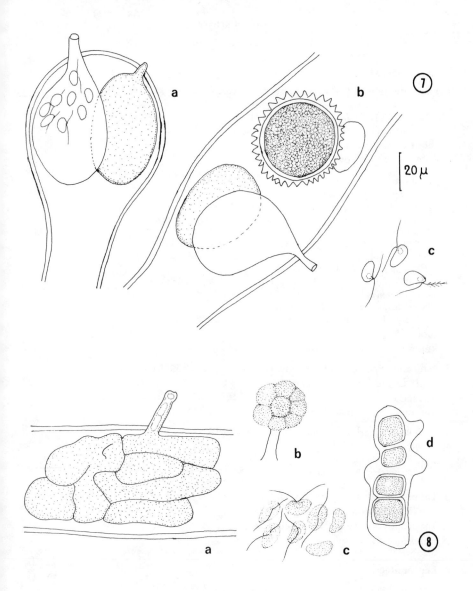

Fig. 7. *Olpidiopsis saprolegniae*, a, sporangia in a swollen hyphal tip; b, oogonium (resting spore) with antheridium; c, zoospores (orig.).

Fig. 8. *Lagenidium destruens*, a, sporangia in a host hypha with a discharge tube; b, formation of zoospores on a discharge tube; c, zoospores; d, small thallus containing resting spores (from Sparrow, 1960).

List of genera

(General reference: Sparrow, 1960)

a. **Lagenidiaceae**

1. **Lagenidium** Schenk - Verh. phys. med. Ges. Würzburg *9*: 27 (1859)
   L. rabenhorstii Zopf (1878, neotype)
   = *Resticularia* Dangeard - Botaniste *2*: 96 (1890/91)
     R. nodosa Dangeard = *L. nodosum* (Dangeard) Ingold
   about 25 species (Fig. 8)
   References: Sparrow, 1960.

2. **Olpidiopsis** Cornu - Ann. Sci. nat., Bot., Sér. 5, *15*: 114 (1872)
   O. saprolegniae (Braun) Cornu (Fig. 7)
   = *Bicilium* Petersen - Ann. mycol. *8*: 503 (1910)
     B. andreei (Lagerh.) Petersen = *O. andreei* (Lagerh.) Sparrow
   = *Diplophysa* Schroet. - Kryptog. Fl. Schles. *3*(1): 195 (1885)
     D. saprolegniae (Cornu) Schroet. = *O. saprolegniae* (Braun) Cornu
   = *Pleocystidium* Fisch - Sber. phys.-med. Soc. Erlangen *16*: 66 (1884)
     P. parasiticum Fisch = *O. schenkiana* Zopf
   = *Pseudolpidiopsis* Minden - Kryptog. Fl. Brandenburg *5*: 255 (1915)
     P. schenkiana (Zopf) Minden = *O. schenkiana* Zopf
   = *Pseudolpidium* Fischer - Rabenh. Krypt. Fl. 1, *4*: 33 (1892)
     P. saprolegniae (Braun) Fischer = *O. saprolegniae* (Braun) Cornu
   about 35 species
   References: Sparrow, 1960.

3. **Petersenia** Sparrow - Dansk bot. Ark. *8* (6): 13 (1934)
   P. lobata (Petersen) Sparrow
   3 species
   References: Sparrow, 1960.

4. **Rozellopsis** Karling - Mycologia *34*: 205 (1942)
   R. septigena Karling
   4 species, all parasitic on Oomycetes
   References: Sparrow, 1960.

5. **Thraustochytrium** Sparrow - Biol. Bull. *70*: 259 (1936)
   T. proliferum Sparrow
   1 saprophytic species found on algae
   References: Sparrow, 1960.

b. **Leptomitaceae**

6. **Apodachlya** Pringsheim - Ber. dt. bot. Ges. *1*: 289 (1883)
   A. brachynema (Hildebr.) Pringsheim
   5 species
   References: Sparrow, 1960.

7. **Apodachlyella** Indoh - Sci. Rep. Tokyo Bunrika Daigaku, sect. B. *4*: 45 (1939)
   A. completa (Humphrey) Indoh
   1 species
   References: Sparrow, 1960.

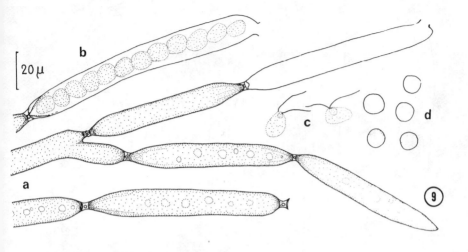

Fig. 9. *Leptomitus lacteus*, a, hyphal segments; b, sporangium; c, primary zoospores; d, encysted zoospores and empty cysts (orig.).

8. **Araiospora** Thaxter - Bot. Gaz. *21*: 326 (1896)
    A. spinosa (Cornu) Thaxt.
5 species
References: Sparrow, 1960.

9. **Leptomitus** Agardh - Systema algarum p. 47 (1824)
    L. lacteus (Roth) Agardh (Fig. 9)
= *Apodya* Cornu - Ann. Sci. nat., Bot. Sér. 5, *15*: 14 (1872)
    A. lactea (Roth) Cornu = *L. lacteus*
1 species in sewage
References: Sparrow, 1960.

10. **Rhipidium** Cornu - Bull. Soc. bot. Fr. *18*: 58 (1871)
    R. interruptum Cornu
7 or 8 species
References: Sparrow, 1960.

11. **Sapromyces** Fritsch - Öst. bot. Z. *43*: 420 (1893)
    S. elongatus (Cornu) Coker
= *Naegeliella* Schroet. in Engl. & Prantl - Nat. Pflanzenfam. *1* (1): 103 (1893) non Correns (1892)
    N. reinschii Schroet. = *S. elongatus*
4 species
References: Sparrow, 1960.

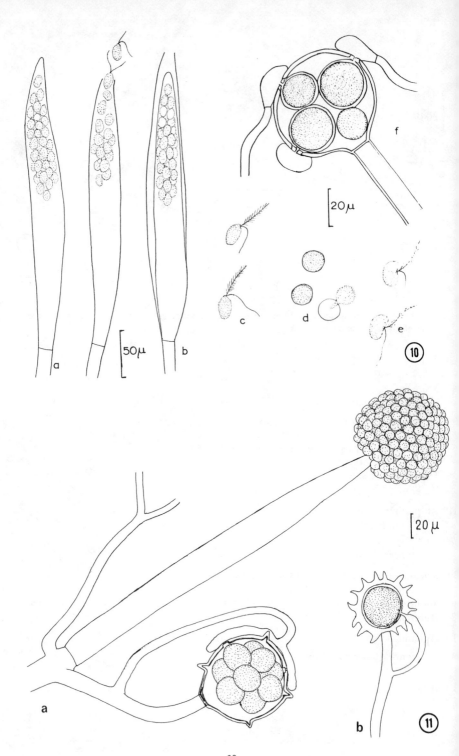

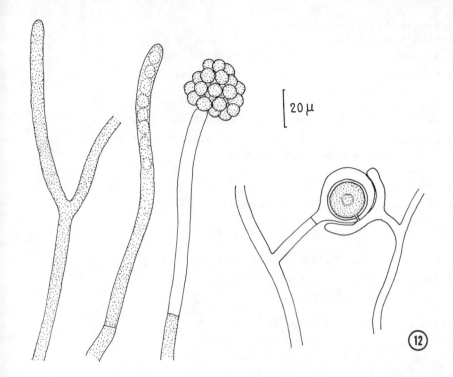

Fig. 10. *Saprolegnia anisospora* de Bary, a, b, zoosporangia, in b proliferating; c, primary zoospores; d, encysting zoospores; e, secondary zoospores; f, oogonium with antheridia and eggs (orig.).
Fig. 11. a, *Achlya oligacantha* de Bary, sporangium with spore cluster and oogonium with antheridium; b, *Achlya radiosa* Maurizio, oogonium with a single egg (orig.).
Fig. 12. *Aphanomyces laevis*, hyphal tips, young sporangium, sporangium with spore cluster and oogonium with antheridium (orig.).

c. **Saprolegniaceae**

12. **Achlya** Nees - Nova Acta Acad. Leop.-Carol. *11*: 514 (1823)
    A. prolifera Nees
    = *Hydronema* Carus - Nova Acta Acad. Leop.-Carol. *11*: 491 (1823) (nomen nudum)
    = *Protoachlya* Coker - Sprolegniaceae p. 90 (1923)
    P. paradoxa Coker
    about 40 species (Fig. 11)
    References: Johnson, 1956.

13. **Aphanodictyon** Huneycull - J. Elisha Mitchell scient. Soc. *64*: 279 (1948)
    A. papillatum Huneycull
    1 species

14. **Aphanomyces** de Bary - Jb. wiss. Bot. *2*: 178 (1860)
    A. stellatus de Bary
    about 10 species parasitic on algae, higher plants and arthropods (Fig. 12)
    References: Coker, 1923; Cutter, 1941; Prowse, 1954; Ziegler, 1958; Scott, 1961.

15. **Aplanes** de Bary - Bot. Ztg *46*: 613 (1888)
    A. braunii de Bary
    The genus is quite doubtful according to Johnson (1956).

16. **Brevilegnia** Coker & Couch - J. Elisha Mitchell scient. Soc. *42*: 212 (1927)
    B. subclavata Couch
    about 10 species
    References: Rossy-Vanderrama, 1956; Höhnk, 1952; Johnson, 1956.

17. **Calyptralegnia** Coker - J. Elisha Mitchell scient. Soc. *42*: 219 (1927)
    C. achlyoides (Coker & Couch) Coker
    2 species
    References: Höhnk, 1953.

18. **Cladolegnia** Johannes - Feddes Repertorium *58*: 211 (1955)
    C. intermedia (Coker) Johannes
    (= *Isoachlya* p.p.) (= *Saprolegnia* sensu Seymour, 1970)
    about 10 species
    References: Johannes, 1955; Johnson and Surratt, 1955.

19. **Dictyuchus** Leitgeb - Bot. Ztg *26*: 503 (1868)
    D. monosporus Leitgeb
    5 species
    References: Coker and Matthews, 1937.

20. **Leptolegnia** de Bary - Bot. Ztg *46*: 609 (1888)
    L. caudata de Bary
    3 species
    References: Coker and Matthews, 1937.

21. **Pythiopsis** de Bary - Bot. Ztg *46*: 609 (1888)
    P. cymosa de Bary
    3 species
    References: Coker, 1923; Coker and Matthews, 1937

22. **Saprolegnia** Nees - Nova Acta Leop.-Carol. *11*: 513 (1823)
    S. ferax (Gruith.) Thuret
    = *Diplanes* Leitgeb - Jb. wiss. Bot. 7: 374 (1869)
      D. saprolegnioides Leitgeb = *S. monoica* Pringsheim
    = *Isoachlya* Kauffman - Am. J. Bot. *8*: 231 (1921)
      I. toruloides Kauffman & Coker = *S. torulosa* de Bary
    about 15 species, *S. parasitica* Coker is parasitic on fish and other aquatic animals
    (Fig. 10)
    References: Coker, 1923; Coker and Matthews, 1937, Seymour, 1970.

23. **Thraustotheca** Humphrey - Trans. Am. phil. Soc. *17*: 131 (1893)
    T. clavata (de Bary) Humphrey
    2 or 3 species
    References: Coker, 1923.

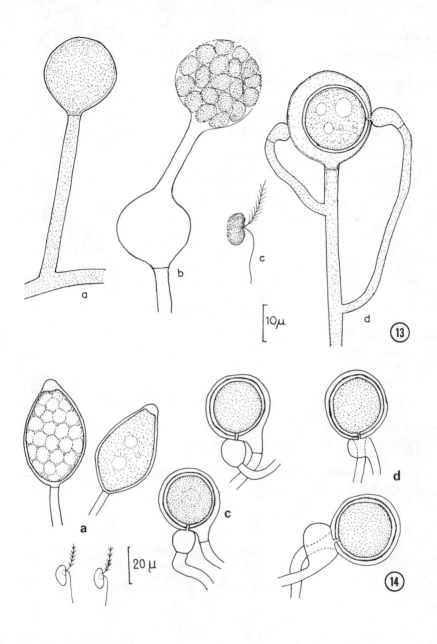

Fig. 13. *Pythium debaryanum* Hesse, a, young sporangium; b, sporangium germinating with a vesicle; c, zoospore; d, oogonium with two antheridia and an egg cell (orig.).
Fig. 14. a, b, c, *Phytophthora cactorum*, a, sporangia; b, zoospores; c, oogonia with antheridia; d, *Phytophthora parasitica*, oogonia with antheridia (orig.).

d. Peronosporaceae

24. **Phytophthora** de Bary - Jl R. agric. Soc. 2, *12*: 240 (1876)
    P. infestans (Mont.) de Bary
= *Blepharospora* Petri - R. C. Accad. Lincei 5, *26*: 299 (1917)
    B. cambivora Petri = *P. cambivora* (Petri) Buisman
= *Kawakamia* Miyabe - Bot. Mag. Tokyo *17*: 306 (1903)
    K. cyperi (Miyabe & Ideta) Miyabe = *P. cyperi* (Ideta) Ito
= *Mycelophagus* Mangin - C. R. Séanc. hebd. Acad. Sci., Paris *136*: 472 (1903)
    M. castaneae Mangin
= *Nozemia* Pethybr. - Sci. Proc. R. Dublin Soc. *13*: 556 (1913)
    N. cactorum (Leb. & Cohn) Pethybr. = *P. cactorum* (Leb. & Cohn) Schroet.
= *Phloeophthora* Klebahn - Zentbl. Bakt. Parasitkde, Abt. 2, *15*: 336 (1906)
    P. syringae Klebahn = *P. syringae* (Klebahn) Klebahn
= *Pseudopythium* Sideris - Phytopathology *20*: 953 (1930)
    P. phytophthoron Sideris = *P. cinnamomi* Rands
= *Pythiacystis* R. E. & E. H. Smith - Bot. Gaz. *42*: 221 (1906)
    P. citrophthora Smith = *P. citrophthora* (Smith) Leonian
= *Pythiomorpha* Petersen - Bot. Tidsskr. *29*: 391 (1909)
    P. gonapodyides Petersen = *P. gonapodyides* (Petersen) Buisman
about 20 species in soil and water, some are important parasites of higher plants (Fig. 14)
References: Tucker, 1931 (Reprint 1967); Waterhouse, 1956, 1963.

25. **Pythiogeton** Minden in Falck - Mycol. Unters. *2* (2): 241 (1916)
    P. ramosus Minden
1 species
References: Sparrow, 1960.

26. **Pythium** Pringsheim - Jb. wiss. Bot. *1*: 304 (1858) (nom. cons.)
    P. monospermum Pringsheim
synonyms vide Waterhouse, 1968
about 70 species in soil and water, often parasitic on roots of higher plants (Fig. 13)
References: Matthews, 1931; Middleton, 1943; Waterhouse, 1967, 1968; van der Plaats-Niterink, 1968.

MUCORALES

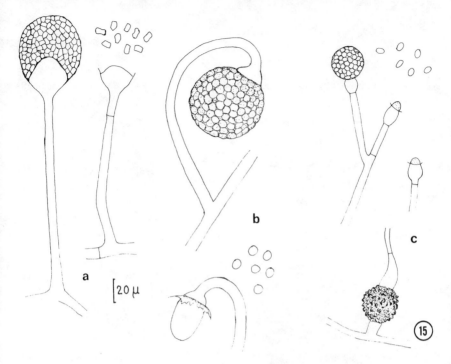

Fig. 15. a, *Absidia cylindrospora* Hagem, sporangia; b, *Circinella mucoroides*, sporangia; c, *Gongronella butleri*, sporangia and zygospore (orig.).

List of genera

(General reference: Zycha, Siepmann and Linnemann, 1969)

a. **Mucoraceae**

1. **Absidia** van Tieghem - Ann. Sci. nat., Bot., Ser. 6, *4*: 350 (1876)
   A. reflexa van Tieghem
   = *Tieghemella* Berl. & de Toni - Syll. Fung. 7: 215 (1888)
   A. repens van Tieghem
   = *Lichtheimia* Vuill. - Bull. Soc. mycol. Fr. *19*: 124 (1903)
   L. corymbifera Vuill. = *A. lichtheimii* (Luc. & Cost.) Lendn.
   = *Mycocladus* Beauverie - Ann. Univ. Lyon n.s., *1*: 163 (1900)
   M. verticillatus Beauv. = *A. verticillata* (Beauv.) Lendn.
   = *Proabsidia* Vuill. - Bull. Soc. mycol. Fr. *19*: 116 (1903)
   P. saccardoi (Oud.) Vuill. = *A. coerulea* Bain.
   = *Pseudoabsidia* Bain. - Bull. Soc. mycol. Fr. *19*: 153 (1903)
   P. vulgaris Bain.
   = *Chlamydoabsidia* Hesseltine & Ellis - Mycologia *58*: 761 (1966)
   Ch. padeni Hesseltine & Ellis = *A. padeni* (Hesseltine & Ellis) Milko
   about 20 species (Fig. 15a)
   References: Hesseltine and Ellis, 1961, 1964, Ellis and Hesseltine 1965, 1966;
   Milko, 1970.

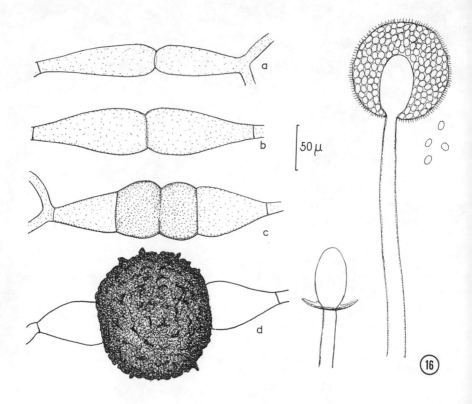

Fig. 16. *Mucor mucedo*, a-d, zygospore formation; f. sporangia with columella (orig.).

2. **Actinomucor** Schostak. - Ber. dt. bot. Ges. *16*: 155 (1898)
    A. repens Schostak. = *A. elegans* (Eidam) Benj. & Hesseltine
= *Glomerula* Bain. - Bull. Soc. mycol. Fr. *19*: 154 (1903)
    G. repens Bain. = *A. elegans*
1 species
References: Benjamin and Hesseltine, 1957.

3. **Circinella** van Tieghem & Le Monnier - Ann. Sci. nat., Bot., Sér. 5, *17*: 298 (1873)
    C. umbellata van Tieghem & Le Monnier
8 species (Fig. 15b)
References: Hesseltine and Fennell, 1955.

4. **Dicranophora** Schroet. - Jber. schles. Ges. vaterl. Kultur *64*: 184 (1886)
    D. fulva Schroet.
1 species
References: Dobbs, 1938.

5. **Gilbertella** Hesseltine - Bull. Torrey bot. Club *87*: 24 (1960)
    G. persicaria (Eddy) Hesseltine
1 species

6. **Gongronella** Ribaldi - Riv. Biol. gen., N.S. *44*: 164 (1952)
   G. urceolifera Ribaldi = *G. butleri* (Lendn.) Peyronel & Dal Vesco (Fig. 15c)
   second species: *C. lacrispora* Hesseltine & Ellis
   References: Hesseltine and Ellis, 1964.

7. **Mucor** Mich. ex Fr. - Syst. mycol. *3*: 317 (1832)
   M. mucedo Fr. (Fig. 16)
   = *Chlamydomucor* Bref. - Unters. Mykol. *8*: 223 (1889)
   Ch. racemosus (Fres.) Bref. = *M. racemosus* Fres.
   and some older names
   about 40 species
   common are: *M. hiemalis* Wehmer, *M. racemosus* Fres., *M. pusillus* Lindt (thermophilic), *M. plumbeus* Bon.
   References: Hesseltine, 1954 (for homothallic species); Schipper, 1967, 1968, 1969, 1973; Baijal and Mehrotra, 1965; Mehrotra and Nand, 1967.

7a. **Backusella** Ellis & Hesseltine - Mycologia *61*: 863 (1969)
   B. circina Ellis & Hesseltine = *Mucor pseudolamprosporus* Naganishi & Hirahara, closely related to *Mucor lamprosporus* Lendner

8. **Parasitella** Bain. - Bull. Soc. mycol. Fr. *19*: 153 (1903)
   P. simplex Bain.
   1 species, parasitic on other Mucorales

9. **Phycomyces** Kunze ex Fr. - Syst. mycol. *3*: 308 (1832)
   P. nitens Kunze ex Fr.
   3 or 4 species
   References: Benjamin and Hesseltine, 1959.

10. **Pirella** Bain. - Ann. Sci. nat., Bot., Sér. 16, *15*: 84 (1883)
    P. circinans Bain.
    1 species
    References: Hesseltine, 1960.

11. **Rhizopodopsis** Boedijn - Sydowia *12*: 330 (1958)
    R. javensis Boedijn
    1 species, not yet cultivated.

12. **Rhizopus** Ehrenb. ex Corda - Icon. Fung. *2*: 20 (1838)
    R. nigricans Ehrenb. = *R. stolonifer* (Ehrenb. ex Fr.) Vuill. (Fig. 17)
    15 or more species; common or important are: *R. oryzae* Went & Prinsen Geerlings, *R. arrhizus* Fischer, *R. delemar* (Boidin) Wehmer & Hanzawa, *R. javanicus* Takeda
    Reference: Inui, Takeda and Iizuka, 1965.

13. **Saksenaea** Saksena - Mycologia *45*: 434 (1953)
    S. vasiformis Saksena
    1 species.

14. **Spinellus** van Tieghem - Ann. Sci. nat., Bot., Sér. 6, *1*: 66 (1875)
    S. fusiger (Link ex Fr.) van Tieghem
    5 species.

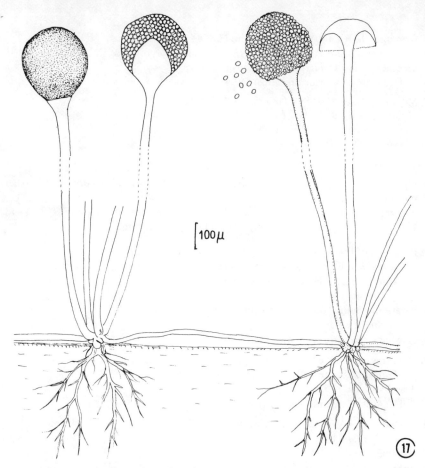

Fig. 17. *Rhizopus stolonifer*, stolons, rhizoids, young, ripe and empty sporangia (orig.).

15. **Sporodiniella** Boedijn - Sydowia *12*: 336 (1958)
     S. umbellata Boedijn
  1 species.

16. **Syzygytes** Ehrenb. ex Fr. - Syst. mycol. *3*: 329 (1832)
     S. megalocarpus Ehrenb. ex Fr. (Fig. 18a)
  = *Sporodinia* Link - Spec. Plant. 6, *1*: 94 (1824)
    S. grandis Link = *S. megalocarpus*
  = *Azygites* Fr. - Syst. mycol. *3*: 330 (1832)
    A. mougeotii Fr. = *S. megalocarpus*
  1 species
  References: Hesseltine, 1957.

17. **Zygorhynchus** Vuill. - Bull. Soc. mycol. Fr. *19*: 116 (1903)
     Z. heterogamus Vuill. (Fig. 18b)
  7 or 8 species
  References: Hesseltine, Benjamin and Mehrotra, 1959; Schipper and Hintikka, 1969.

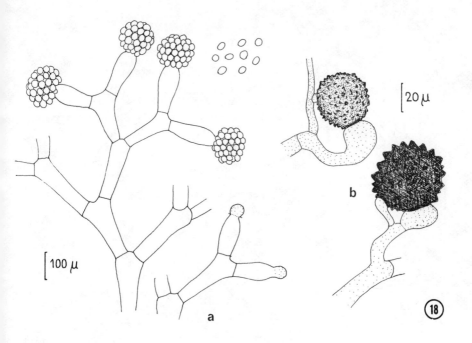

Fig. 18. a, *Syzygytes megalocarpus*, sporangiophores and sporangia; b, *Zygorhynchus heterogamus*, zygotes (orig.).

b. Thamnidiaceae and Choanephoraceae

18. **Choanephora** Currey - J. Linn. Soc. Bot. *13*: 576 (1873)
    Ch. infundibulifera (Currey) Sacc.
    ≡*Cunninghamia* Currey - l.c. p. 334
    ≡*Choanephorella* Vuill. - Bull. Soc. mycol. Fr. *20*: 28 (1904)
    = *Blakeslea* Thaxter - Bot. Gaz. *58*: 353 (1914)
      B. trispora Thaxter = *Ch. trispora* (Thaxter) Sinha
    Further species are *Ch. cucurbitarum* (Berk. & Rav.) Thaxter (Fig. 19c), *Ch. circinans* (Naganishi & Kawakami) Hesseltine & Benjamin and *Ch. conjuncta* Couch
    References: Sinha, 1940; Hesseltine and Benjamin, 1957.

19. **Chaetocladium** Fres. - Beitr. Mykol. p. 97 (1863)
    Ch. jonesii (Berk. & Br.) Fres.
    second species: *Ch. brefeldii* van Tieghem & Le Monnier.
    Both species are parasitic on Mucorales.

20. **Thamnostylum** v. Arx & Upadhyay ap. v. Arx - Gen. Fungi p. 247 (1970)
    T. piriforme (Bain.) v. Arx & Upadhyay = *Helicostylum piriforme* Bain. in Bull. Soc. bot. Fr. *27*: 227 (1880)
    second species: *Thamnostylum lucknowense* (Rai & al.) v. Arx & Upadhyay

43

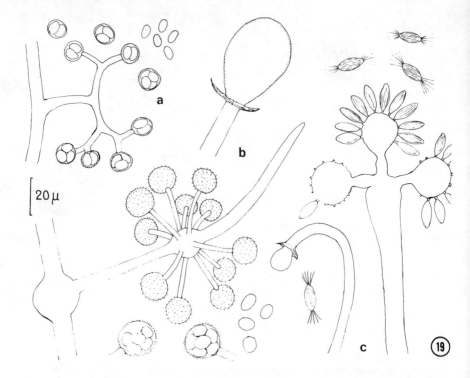

Fig. 19. a, *Thamnidium elegans*, part of a sporangiophore with sporangiola; b, *Helicostylum fresenii*, terminal sporangium with columella and branches of a sporangiophore with sporangiola; c, *Choanephora cucurbitarum*, recurved sporangium, sporangiospores and conidiophore with conidia (orig.).

= *Helicostylum lucknowense* Rai & al. - Can. J. Bot. *39*: 1281 (1961).
References: Upadhyay, 1973.

21. **Cokeromyces** Shanor - Mycologia *42*: 272 (1950)
    C. recurvatus Poitras
    second species: *C. poitrasii* Benjamin - Aliso *4*: 523 (1960).

22. **Helicostylum** Corda - Icon. Fung. *5*: 18. 55 (1842)
    H. elegans Corda
    = *Bulbothamnidium* Klein - Verh. zool.-bot. Ges. Wien *20*: 557 (1870)
        B. elegans Klein = *H. fresenii* (van Tieghem & le Monnier) Lythgoe (Fig. 19b)
    = *Chaetostylum* van Tieghem & le Monnier - Ann. Sci. nat., Bot. Sér. 5, *17*: 328 (1873)
        Ch. fresenii van Tieghem & le Monnier = *H. fresenii*
    6 species
    References: Hesseltine and Anderson, 1957; Lythgoe, 1958; Upadhyay, 1973.

23. **Hesseltinella** Upadhyay - Persoonia *6*: 111 (1970)
    H. vesiculosa Upadhyay
    1 species.

24. **Radiomyces** Embree - Am. J. Bot. *46*: 25 (1959)
    R. spectabilis Embree
  second species: *Radiomyces embreei* Benjamin - Aliso *4*: 526 (1960).

25. **Thamnidium** Link ex Wallr. - Flora crypt. Germ. *4* (2): 324 (1833)
    T. elegans Link ex Wallr. (Fig. 19a)
  4 species
  References: Hesseltine and Anderson, 1956.

c. **Pilobolaceae**

26. **Pilobolus** Tode ex Fr. - Syst. mycol. *2*: 308 (1823)
    P. crystallinus Tode ex Fr.
  9 species
  References: Page, 1956; Nand and Mehrotra, 1968.

27. **Pilaira** van Tieghem - Ann. Sci. nat., Bot., Sér. 6, *1*: 51 (1875)
    P. anomala (Ces.) Schroet.
  4 species, mycelium very scanty.

27a. **Utharomyces** Boedijn - Sydowia *12*: 340 (1958)
    U. epallocaulus Boedijn
  1 species, in warmer areas, with abundant mycelium.

d. **Piptocephalidaceae and Dimargaritaceae**

28. **Dimargaris** van Tieghem - Ann. Sci. nat., Bot., Sér. 6, *1*: 154 (1875)
    D. crystallina van Tieghem
  7 species, parasitic on Mucorales (Fig. 20b)
  References: Benjamin, 1959.

29. **Dispira** van Tieghem - Ann. Sci. nat., Bot., Sér. 6, *1*: 160 (1875)
    D. cornuta van Tieghem
  2 species, parasitic on other Mucorales
  References: Benjamin, 1959.

30. **Piptocephalis** de Bary - Abh. Senckenb. naturf. Ges. *5*: 356 (1865)
    P. freseniana de Bary
  ≡*Mucoricola* Nieuwland - Am. Midl. Nat. *4* 82 (1916)
    P. freseniana de Bary
  10-15 species, all parasitic on other fungi (Mucorales and *Penicillium*)
  References: Benjamin, 1959.

31. **Syncephalastrum** Schroet. - Krypt. Fl. Schles. 3, (2): 217 (1886)
    S. racemosum Cohn ex Schroet. (Fig. 20a)
  1 species
  References: Benjamin, 1959.

32. **Syncephalis** van Tieghem & Le Monnier - Ann. Sci. nat., Bot., Sér. 5, *17*: 372
  (1873)
    S. cordata van Tieghem & Le Monnier
  = *Microcephalis* Bain. - Étude des Mucorinées p. 114 (1882)
    S. curvata Bain.

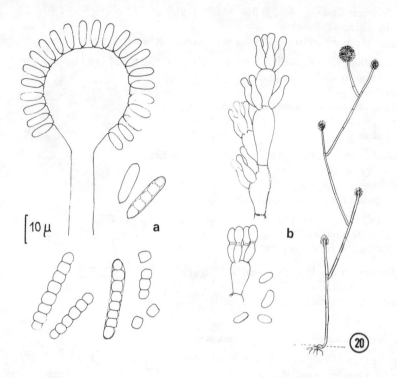

Fig. 20. a, *Syncephalastrum racemosum*, terminal swelling with young merosporangia, merosporangia of different maturity and spores; b, *Dimargaris cristalligena*, habitat of sporophore, terminal portion of sporiferous branchlet and a single branchlet with spores (from Benjamin, 1959).

= *Clavocephalis* Bain. - l.c. p. 120
　　S. nodosa van Tieghem
= *Monocephalis* Bain. - l.c. p. 123
　　S. fusiger Bain.
= *Syncephalopsis* Boedijn - Sydowia *12*: 353 (1958)
　　S. bispora Rac.
= *Syncephalidium* Badura - Allionia *9*: 179 (1963)
　　S. penicillatum Badura
= *Acephalis* Badura & Badurowa - Acta Soc. Bot. Pol. *33*: 519 (1964)
　　A. radiata Badura & Badurowa

20-30 species, parasitic on other fungi, especially on Mucorales (Fig. 21a)
References: Benjamin, 1959, Milko, 1967; Mehrotra and Prasad, 1965, 1967.

33. **Tieghemiomyces** Benjamin - Aliso *4*: 390 (1959)
　　T. californicus Benjamin
second species: *T. parasiticus* Benjamin - Aliso *5*: 11 (1961)
both parasitic on other fungi.

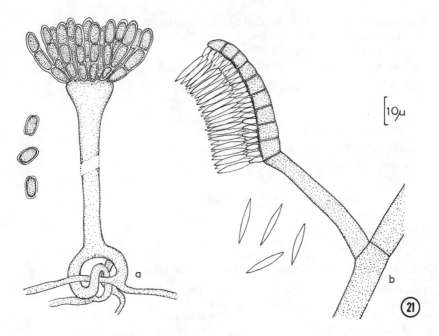

Fig. 21. a, *Syncephalis nodosa* van Tieghem, terminal swelling with merosporangia and spores; b, *Coemansia pectinata* Bain., sporocladium with phialides and spores (from Benjamin, 1959).

e. **Kickxellaceae**

34. **Coemansia** van Tieghem & Le Monnier - Ann. Sci. nat., Bot., Sér. 5, *17*: 392 (1872)
    C. reversa van Tieghem & Le Monnier
    10 or more species (Fig. 21b)
    References: Benjamin, 1958; Linder, 1943.

35. **Dipsacomyces** Benjamin - Aliso *5*: 15 (1961)
    D. acuminosporus Benjamin
    1 species.

36. **Kickxella** Coemans - Bull. Soc. Roy. Bot. Belg. *1*: 155 (1862)
    K. alabastrina Coemans
    = *Coronella* Crouan - Florule du Finistère p. 12 (1867)
       C. nivea Crouan = *K. alabastrina*
    = *Coemansiella* Sacc. - Syll. Fung. *2*: 815 (1883)
       K. alabastrina Coemans
    1 species
    References: Benjamin, 1958; Linder, 1943.

37. **Linderina** Raper & Fennell - Am. J. Bot. *39*: 81 (1952)
    L. pennispora Raper & Fennell
    1 species.

38. **Martensella** Coemans - Bull. Acad. Roy. Sci. Lett. Belg., 2, *15*: 540 (1863)
   M. pectinata Coemans
   second species: *M. corticii* Thaxter ex Linder
   References: Benjamin, 1959; Jackson and Dearden, 1948.

39. **Martensiomyces** Meyer - Bull. Soc. mycol. Fr. *73*: 189 (1957)
   M. pterosporus Meyer
   1 species
   References: Benjamin, 1959.

40. **Spirodactylon** Benjamin - Aliso *4*: 408 (1959)
   S. aureum Benjamin
   1 species.

41. **Spiromyces** Benjamin - Aliso *5*: 273 (1963)
   S. minutus Benjamin
   1 species.

f. Cunninghamellaceae, Mortierellaceae and other families

42. **Cunninghamella** Matr. - Ann. mycol. *1*: 46 (1903)
   C. echinulata (Thaxt.) Thaxt. (Fig. 22a)
   = *Actinocephalum* Saito - Bot. Mag. Tokyo *19*: 36 (1905)
   A. japonicum Saito
   = *Saitomyces* Ricker - J. Mycol. *12*: 61 (1906)
   S. japonicus (Saito) Ricker
   = *Muratella* Bain. & Sart. - Bull. Soc. mycol. Fr. *29*: 129 (1913)
   M. elegans Bain. & Sart.
   6 or 7 species
   References: Cutter, 1946; Milko and Belyakova, 1967; Samson, 1969.

43. **Echinosporangium** Malloch - Mycologia *59*: 327 (1967)
   E. transversale Malloch
   1 species.

44. **Endogone** Link ex Fr. - Syst. mycol. *2*: 295 (1823)
   E. pisiformis Link ex Fr.
   more than 20 species, most of them are mycorrhizal symbionts and do not develop well in pure culture.

45. **Mortierella** Coemans - Bull. Acad. Roy. Sci. Belg. II, *15*: 288 (1863)
   M. polycephala Coemans
   = *Carnoya* Dewevre - Grevillea *22*: 4 (1893)
   C. capitata Dewevre
   = *Haplosporangium* Thaxter - Bot. Gaz. *58*: 362 (1914)
   H. bisporale Thaxter = *M. bisporalis* (Thaxter) Björling
   = *Naumoviella* Novotelnova - Not. syst. Sect. crypt. Inst. bot. Acad. Sci. USSR *6*: 155 (1950)
   N. nivea Novotelnova = *M. candelabrum* van Tieghem & Le Monnier
   about 60 species (Fig. 22b)
   References: Linnemann, 1941, 1958; Gams and Williams, 1963; Björling, 1936; Wolf, 1954; Turner, 1963; Gams, 1969; Domsch and Gams, 1970; Gams & al., 1972.

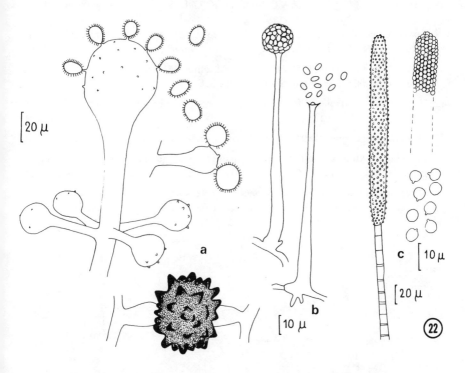

Fig. 22. a, *Cunninghamella echinulata*, branched conidiophores with conidia and zygospore; b, *Mortierella alpina* Peyronel, sporangiophores with sporangiola and sporangiospores; c, *Mycotypha microspora* Fenner, conidiophores and conidia (orig.).

46. **Mycotypha** Fenner - Mycologia *24*: 185 (1932)
    M. microspora Fenner (Fig. 22c)
    second species: *M. africana* Novak & Backus
    References: Novak and Backus, 1963.

47. **Rhopalomyces** Corda - Prachtflora p. 3 (1839)
    R. elegans Corda
    3 or 4 species
    References: Boedijn, 1927, 1958, Ellis 1963.

ENTOMOPHTHORALES

1. Spores campanulate, apex pointed, base truncate, sporophores swollen, without columella, resting spores are azygospores (entomogenous). . . . *Myiophyton* (4)

1. Spores not bell-shaped, mostly spherical, ellipsoidal or pyriform, sporophores often columellate or not swollen. 2

2. Entomogenous (isolated from insects), spores pyriform or ellipsoidal . . . .
. . . . *Entomophthora* (3)

2. Not entomogenous, mostly saprophytic 3

3. Mycelium extensive, often septate, gametangia unequal, conidiophores elongated . . . . *Conidiobolus* (2)

3. Mycelium not extensive, often becoming yeast-like, sporophores short, often conical (isolated from excrements of frogs and lizards) . . . . . . *Basidiobolus* (1)

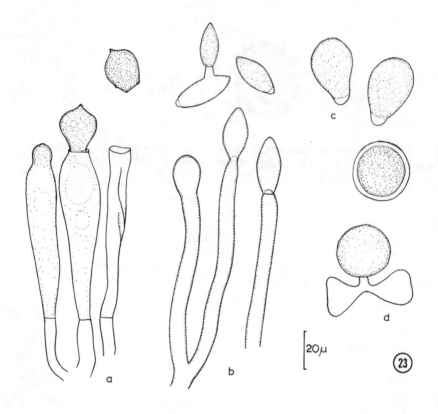

Fig. 23. a, *Myiophyton muscae*, sporophores and spores; b, *Entomophthora aphidis*, sporophores and spores; c, *Entomophthora grylli*, spores; d, *Entomophthora zygospora*, zygospore (a, c, orig., b from Krenner, 1961, d from MacLeod, 1956).

List of genera

1. **Basidiobolus** Eidam in Cohn - Beitr. Biol. Pfl. *4*: 181 (1886)
   B. ranarum Eidam
   1 or a few species
   References: Drechsler, 1947, 1956, 1964.

2. **Conidiobolus** Bref. - Unters. Mykol. *4*: 35 (1884)
   C. utriculosus Brefeld
   about 20 species
   References: Couch, 1939; Drechsler, 1953, 1954, 1960, 1965; Srinivasan and Thirumalachar, 1963, 1967.

3. **Entomophthora** Fres. - Bot. Ztg *14*: 882 (1856)
   E. sphaerosperma Fres. sensu Gustafsson, 1965
   = *Boudierella* Cost. - Bull. Soc. mycol. Fr. *13*: 38 (1897)
   = *Delacroixia* Sacc. & Syd. - Syll. Fung. *14*: 457 (1899)
      D. coronata (Cost.) Sacc. & Syd. = *E. coronata* (Cost.) Kevorkian
   = *Entomophaga* Batko - Bull. Acad. polon. Sci., sér. biol. *12*: 325 (1964)
      E. grylli (Fres.) Batko = *Entomophthora grylli* Fres.
   = *Triplosporium* (Thaxt.) Batko - l.c. p. 324
      T. fresenii (Nowak.) Batko = *Entomophthora fresenii* (Nowak.) Gustafsson
   = *Zoophthora* Batko - l.c. p. 323
      Z. radicans (Bref.) Batko = *Empusa radicans* Bref. = *Entomophthora sphaerosperma* Fres.
   about 30 species (Fig. 23b-d)
   References: Lakon, 1963; Gustafsson, 1965.

4. **Myiophyton** Lebert - Neue Denkschr. schweiz. Ges. Naturw. *15*: (1856)
   M. cohnii Lebert = *M. muscae* (Cohn) Krenner (Fig. 23a)
      Basionym: *Empusa muscae* Cohn — Nova Acta Leop. *25*: 301 (1855)
   ≡ *Empusa* Cohn - Nova Acta Leop. *25*: 301 (1855) non *Empusa* Lindley
   = *Lamia* Nowak. - Pam. Akad. Krakau *8*: 153 (1884)
      L. culicis (Braun) Nowak. = *Empusa culicis* Braun
      = *Myiophyton culicis* (Braun) v. Arx
   ≡ *Culicicola* Nieuwl. - Am. Midl. Nat. *4*: 378 (1916)
      C. culicis (Braun) Nieuwl. = *M. culicis*
   Other species: *M. planchonianum* (Cornu) v. Arx
   5 or 6 species
   References: Krenner, 1961; Gustafsson, 1965.

ENDOMYCETALES

15. Septate hyphae 3-6 μm wide, septa thickened around a central plugged pore (visible under the light microscope) 16

15. Septate hyphae 2-4 μm wide, without plugged septal pores 17

16. Nitrate assimilated, ascospores smaller than 5 μm, asci catenulate. . . . . *Hormoascus* (6)

16. Nitrate not assimilated, ascospores larger . . . . *Ambrosiozyma* (5)

17. Budding cells absent, conidia with a truncate base borne on elongating conidiogenous cells present, asci borne laterally on hyphae, often in chains . . . . *Botryoascus* (7)

17. Budding cells present. 21

18. Ascospores lunate, small, with apical caps . . . . *Guilliermondella* (32)

18. Ascospores needle-shaped or elongated-fusiform 19

19. Budding cells absent, asci are hyphal cells, often in chains, many-spored . . . . *Eremothecium* (29)

19. Budding cells present . . . . *Nematospora* (30)

20. Ascospores needle-shaped . . . . *Metschnikowia* (31)

20. Ascospores not needle-shaped 21

21. Special cultivation conditions necessary (temp. 35-40° and low pH). . . . . *Cyniclomyces* (12)

21. Special cultivation conditions unnecessary 22

22. Strong acid production on sugar-containing media . . . . *Dekkera* (13)

22. Acid production absent or weak 23

23. Ascus thick-walled, with a beak . . . . *Pachysolen* (14)

23. Ascus not thick-walled, no beak. 24

24. Ascus is an appendage on the yeast cell . . . . *Lipomyces* (15)

24. Yeast cell itself becomes an ascus 25

25. Wall of ascus evanescent, ascospores freed at an early stage 26

25. Wall of ascus persistent 28

26. Nitrate assimilated, ascospores lunate, hat-shaped or saturnoid. . . . . *Hansenula* (17)*

26. Nitrate not assimilated 27

27. Ascospores spherical, hat-shaped or saturnoid . . . . *Pichia* (16)*

---

* *Hansenula, Pichia* (and *Williopsis*) are distinguished by Kudrjawzew (1950) on the following morphological characters:

Ascospores hemispherical . . . . *Pichia*
Ascospores hat-shaped . . . . *Hansenula*
Ascospores saturnoid . . . . *Williopsis*

Such a classification seems to be a more natural one.

27. Ascospores prolate ellipsoidal or reniform .... *Kluyveromyces* (18)

28. Ascospores smooth 29

28. Ascospores rugose 31

29. Ascospores spherical, usually 1-4 per ascus .... *Saccharomyces* (21)

29. Ascospores not spherical................................... 30

30. Ascospores large, prolate ellipsoidal, 1 or rarely 2 per ascus ....
   ....*Lodderomyces* (19)

30. Ascospores oblate ellipsoid or lentiform, 1-4 per ascus .... *Wingea* (20)

31. Nitrate assimilated, ascospores spheroidal, verrucose ..... *Citeromyces* (22)

31. Nitrate not assimilated....................................... 32

32. Ascospores saturn-shaped, verrucose, usually 1 per ascus ....
   ....*Schwanniomyces* (23)

32. Ascospores spherical or oval, usually 1-2(-4) per ascus .. *Debaryomyces* (24)

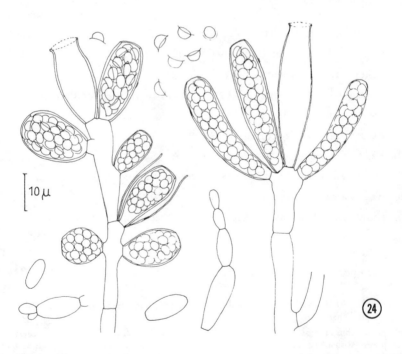

Fig. 24. *Ascoidea corymbosa* W. Gams & Grinbergs, ascophores, asci, partly proliferating, ascospores and budding cells (orig.).

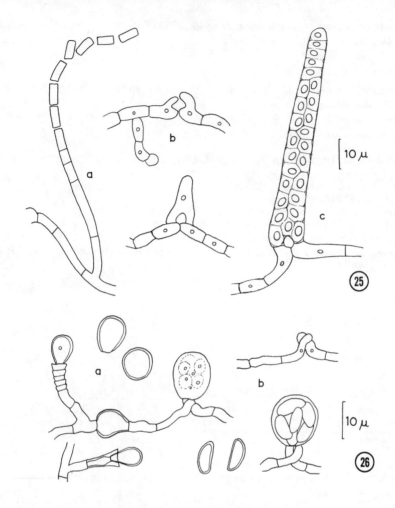

Fig. 25. *Dipodascus aggregatus* Francke-Grosmann, a, hyphae forming fission cells; b, gametangia, forming an ascus; c, ascus (orig.).
Fig. 26. *Eremascus fertilis*, a, conidiophores and conidia; b, gametangia, asci and ascospores (orig.).

List of genera

(General reference: Lodder, 1970)

1. **Ascoidea** Brefeld - Unters. Mykol. *9*: 91 (1891)
   A. rubescens Brefeld
   5 or 6 species (Fig. 24)
   References: Walker, 1931; Batra and Francke-Grosmann, 1964.

2. **Dipodascus** Lagerh. - Jb. wiss. Bot. *24*: 549 (1892)
    D. albidus Lagerh.
  3 or 4 species (Fig. 25)
  References: Batra, 1959.

3. **Eremascus** Eidam in Cohn - Beitr. Biol. Pfl. *3*: 385 (1883)
    E. albus Eidam
  second species: *E. fertilis* Stoppel (Fig. 26)
  References: Harrold, 1950; von Arx, 1967.

4. **Cephaloascus** Hanawa - Jap. Z. Dermatol. Urol. Tokyo *20*: 14 (1920)
    C. fragrans Hanawa (Fig. 27)
  = *Ascocybe* Wells - Mycologia *46*: 37 (1954)
    A. grovesii Wells = *C. fragrans*
  = *Aureomyces* Ruokola & Salonen - Mycopath. Mycol. appl. *42*: 273 (1970)
    A. mirabilis Ruokola & Salonen = *C. fragrans*
  1 species. The genus is related to *Ophiostoma*.
  References: Dixon, 1959; Wilson, 1961; Schippers-Lammertse and Heyting, 1962;
  von Arx, 1967.

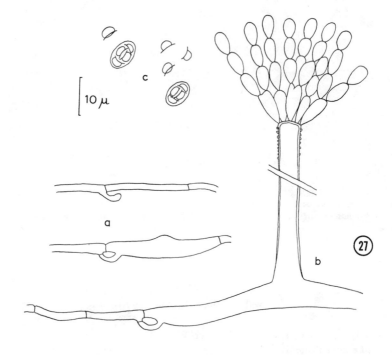

Fig. 27. *Cephaloascus fragrans*, a, formation of a clamp connection; b, ascophore with young asci; c, asci and ascospores (orig.).

56

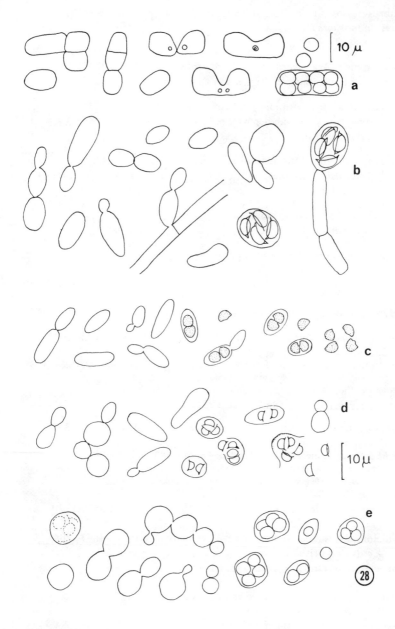

Fig. 28. a, *Schizosaccharomyces octosporus*, fission cells, conjugating cells and formation of endospores (ascospores); b, *Saccharomycopsis fibuliger*; c, *Pichia membranifaciens*; d, *Hansenula anomala*; e, *Saccharomyces cerevisiae*, budding cells and endospores (a, b, e orig., c, d after Kudriawzew, 1960).

5. **Ambrosiozyma** van der Walt - Mycopath. Mycol. appl. *46*: 305 (1972)
   A. monospora (Saito) van der Walt
   2 or 3 species.

6. **Hormoascus** v. Arx - Antonie van Leeuwenhoek *38*: 302 (1972)
   H. platypodis (Baker & Kreger-van Rij) v. Arx

7. **Botryoascus** v. Arx - Antonie van Leeuwenhoek *38*: 304 (1972)
   B. synnaedendrus (van der Walt & Scott) v. Arx
   conidial state: *Raffaelea*.

8. **Arthroascus** v. Arx - Antonie van Leeuwenhoek *38*: 304 (1972)
   A. javanensis (Kloecker) v. Arx (= *Endomyces javanensis* Kloecker), related to
   Hanseniaspora.

9. **Endomyces** Reess - Bot. Unters. p. 77 (1870)
   E. decipiens (Tul.) Reess
   = *Magnusiomyces* Zender - Bull. Soc. bot. Genève *17*: 41 (1925)
   M. ludwigii Zender = *E. magnusii* Ludwig
   conidial state: *Geotrichum*
   6 species; *E. geotrichum* Butler & Petersen is the ascigerous state of *Geotrichum candidum*
   References: Butler and Petersen, 1972; von Arx, 1972.

10. **Saccharomycopsis** Schiönning - C. r. Trav. Lab. Carlsberg *6*: 103 (1903)
    S. capsularis Schiönning (Fig. 31a)
    ≡ *Endomycopsis* Dekker - Verh. K. ned. Akad. Wet., afd. Natuurk., *28*: 264 (1931)
    ≡ *Prosaccharomyces* Novák & Zsolt - Acta bot. hung. 7: 93 (1961)
    5 species (Fig. 28b)
    References: van der Walt and Scott, 1971.

11. **Schizosaccharomyces** Lindner - Wochenschr. Brauerei *10*: 1298 (1893)
    S. pombe Lindner
    ≡ *Schizosaccharis* Clem. & Shear - Gen. Fungi p. 246 (1931)
    = *Octosporomyces* Kudrjawzew - Syst. Hefen p. 247 (1960)
    O. octosporus (Beijerinck) Kudrjawzew = *S. octosporus* Beijerinck
    3 species (Fig. 28a)
    References: Lodder and Kreger-van Rij, 1952; Slooff in Lodder, 1970.

12. **Cyniclomyces** van der Walt & Scott - Mycopath. Mycol. appl. *43*: 284 (1971)
    C. guttulatus (Robin) van der Walt & Scott
    1 species, in intestinal tracts of rabbits.

13. **Dekkera** van der Walt - Antonie van Leeuwenhoek *30*: 274 (1964)
    D. bruxellensis van der Walt
    second species: *D. intermedia* van der Walt
    Dekkera is the ascosporogenous state of *Brettanomyces* (q.v.).

14. **Pachysolen** Boidin & Adzet - Bull. Soc. mycol. Fr. *73*: 339 (1957)
    P. tannophilus Boidin & Adzet
    1 species.

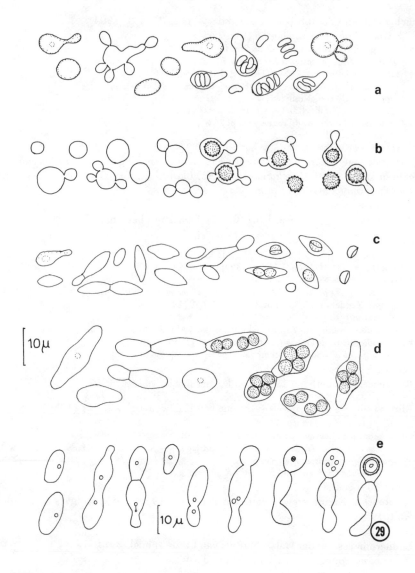

Fig. 29. a, *Kluyveromyces marxianus*; b, *Debaryomyces hansenii*; c, *Hanseniaspora apiculata*; d, *Saccharomycodes ludwigii*; e, *Nadsonia elongata*, formation of budding cells and endospores (after Kudrjawzew, 1960).

15. **Lipomyces** Lodder & Kreger-van Rij - The Yeasts p. 333, 669 (1952)
    L. starkeyi Lodder & Kreger-van Rij
    second species: *L. lipoferus* (den Dooren de Jong) Lodder & Kreger-van Rij.

16. **Pichia** Hansen - Zentbl. Bakt. Parasitkde, Abt. 2, *12*: 534 (1904)
    P. membranefaciens Hansen (Fig. 28c)
= *Petasospora* Boidin & Abadie - Bull. Soc. mycol. Fr. *70*: 353 (1954)
    P. rhodanensis (Ramirez & Boidin) Phaff
= *Issatchenkia* Kudrjawzew - Syst. Hefen p. 161 (1954/1960)
    I. orientalis Kudrjawzew = *P. kudriavzevii* Boidin, Pignal et Besson
= *Zygopichia* (Klöcker) Kudrjawzew - l.c. p. 192
    Z. chevalieri (Guill.) Klöcker = *P. membranefaciens*
= *Zygowillia* (Klöcker) Kudrjawzew - l.c. p. 204
    P. pastoris (Guill.) Phaff
= *Zymopichia* Novák & Zsolt - Acta bot. hung. 7: 99 (1961)
The following generic names are used for haploid mating types of *Pichia guilliermondii* Wickerham: *Myceloblastanon* Ota, *Blastodendrion* (Ota) Cif. & Red., and *Microanthomyces* Grüss (cf. *Candida*)
about 40 species
References: Kreger-van Rij, 1964; Boidin, Pignal and Besson, 1965.

17. **Hansenula** Syd. - Ann. mycol. *17*: 44 (1919)
    H. anomala (Hans.) Syd. (Fig. 28d)
≡ *Willia* Hansen (1904) [non C. Müller 1899]
= *Azymohansenula* Novák & Zsolt - Acta bot. hung. 7: 99 (1961)
    A. canadensis (Wickerham) Novák & Zsolt = *H. canadensis* Wickerham
= *Williopsis* Zender - Bull. Soc. bot. Genève *17*: 258 (1925)
    H. saturnus (Klöcker) Syd.
= *Zygohansenula* Lodder - Zentbl. Bakt. Parasitkde, Abt. 2, *86*: 227 (1932)
= *Zygowilliopsis* Kudrjawzew - Syst. Hefen p. 211 (1960)
    H. californica (Lodder) Wickerham
about 25 species
References: Wickerham, 1951; Lodder and Kreger-van Rij, 1952.

18. **Kluyveromyces** van der Walt - Antonie van Leeuwenhoek *22*: 265 (1956)
    K. polysporus van der Walt
= *Fabospora* Kudrjawzew - Syst. Hefen p. 178 (1960)
    F. macedoniensis (Diddens & Lodder) Kudrjawzew = *K. marxianus* (Hansen) van der Walt
= *Zygofabospora* Kudrjawzew - l.c. p. 183
    Z. marxiana (Hansen) Kudrjawzew = *K. marxianus*
about 20 species (Fig. 29a)
References: Kudrjawzew, 1960; van der Walt, 1956, 1965; Roberts & van der Walt, 1959.

19. **Lodderomyces** van der Walt - Antonie van Leeuwenhoek *32*: 1 (1966)
    L. elongisporus (Recca & Mrak) van der Walt
1 species.

20. **Wingea** van der Walt - Antonie van Leeuwenhoek *33*: 97 (1967)
    W. robertsii (van der Walt) van der Walt
1 species.

21. **Saccharomyces** Meyen ex Hansen - Medd. Carlsberg Lab. *2*: 29 (1883)
    S. cerevisiae Meyen ex Hansen (Fig. 28e)
= *Zygosaccharomyces* Barker - Phil. Trans. R. Soc. Ser. B, *194*: 467 (1901)
    Z. barkeri Sacc. & Syd. = *Saccharomyces rouxii* Boutroux
= *Torulaspora* Lindner - Jb. Versuchs- u. Lehranst. Brau. Berlin 7: 441 (1904)
    T. delbrueckii Lindner = *Saccharomyces delbrueckii* Lindner

= *Zymodebaryomyces* Novák & Zsolt - Acta bot. hung. 7:98 (1961) (pro maxima parte)
about 40 species
References: Lodder and Kreger-van Rij, 1952.

22. **Citeromyces** Santa Maria - Ann. Inst. Nac. Invest. Gar. *17*: 269 (1957)
   C. matritensis Santa Maria
   1 species.

23. **Schwanniomyces** Klöcker - C. r. Trav. Lab. Carlsberg 7: 273 (1909)
   S. occidentalis Klöcker
   3 species
   References: Lodder and Kreger-van Rij, 1952.

24. **Debaryomyces** Klöcker - C. r. Trav. Lab. Carlsberg 7: 273 (1909) sensu Lodder &
   Kreger-van Rij, 1952
   D. hansenii (Zopf) Lodder & Kreger-van Rij (Fig. 29b)
   9 species
   References: Lodder and Kreger-van Rij, 1952.

25. **Hanseniaspora** Zikes - Zentbl. Bakt. Parasitkde, Abt. 2, *30*: 145 (1911)
   H. apiculata (Reess) Zikes = *H. valbyensis* Klöcker (Fig. 29c)
   ≡ *Hansenia* Lindner (1905) non Zopf (1883)
   = *Kloeckerospora* Niehaus - Zentbl. Bakt. Parasitkde, Abt. 2, *86*: 253 (1932)
   K. uvarum Niehaus = *H. uvarum* (Niehaus) Shehata & al.
   = *Vanderwaltia* Novák & Zsolt - Acta bot. hung. 7:98 (1961)
   V. vineae (v. d. Walt & Tscheuschner) Novák & Zsolt = *H. vineae* v. d. Walt & Tscheuschner
   3 species, *Hanseniaspora* is the ascosporogenic state of *Kloeckera* Janke
   References: Kudrjawzew, 1960; Kreger-van Rij and Ahearn, 1968; Miller and
   Phaff, 1958.

26. **Saccharomycodes** Hansen - Zentbl. Bakt. Parasitkde, Abt. 2, *5*:632 (1889)
   S. ludwigii Hansen (Fig. 29d)
   = *Saënkia* Kudrjawzew - Syst. Hefen p. 255 (1960)
   S. bispora (Castelli) Kudrjawzew = *S. ludwigii*
   1 species
   References: Kudrjawzew, 1960; Miller and Phaff, 1958.

27. **Nadsonia** Syd. - Ann. mycol. *10*: 347 (1912)
   N. fulvescens (Nadson & Konokotina) Syd.
   ≡ *Guilliermondia* Nadson & Konokotina (1911) non Boud.
   2 species (Fig. 29e)
   References: Lodder and Kreger-van Rij, 1952.

28. **Wickerhamia** Soneda - Nagaoa 7: 9 (1960)
   W. fluorescens Soneda
   1 species.

29. **Eremothecium** Borzi - Boll. Soc. bot. ital. *20*: 455 (1888)
   E. cymbalariae Borzi
   = *Crebrothecium* Routien - Mycologia *41*: 184 (1949)
   C. ashbyi (Guilliermond) Routien = *E. ashbyi* Guill.
   2 or 3 species
   References: Krneta-Jordi, 1962; Guilliermond, 1936.

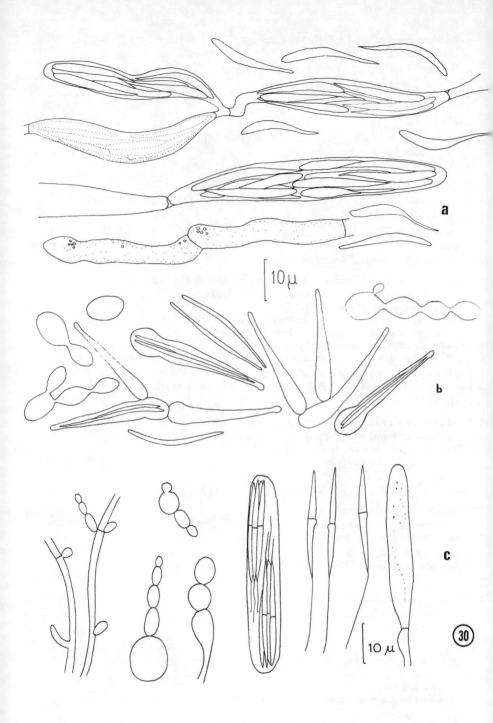

Fig. 30. a, *Eremothecium ashbyi*; b, *Metschnikowia bicuspidata*; c, *Nematospora coryli* (orig.).

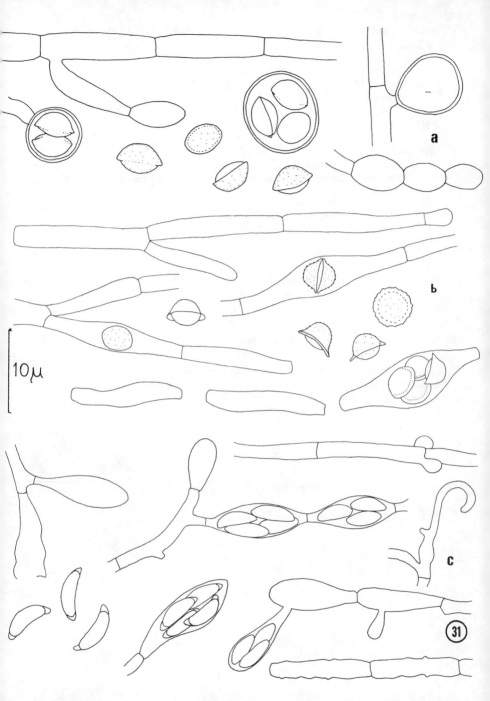

Fig. 31. a, *Saccharomycopsis capsularis*; b, *Arthroascus javanensis*; c, *Guilliermondella seleno-spora* (orig.).

27. **Nematospora** Peglion - Zentbl. Bakt. Parasitkde, Abt. 2, *7*: 754 (1901)
   N. coryli Peglion (Fig. 30a-c)
   = *Ashbya* Guilliermond - Rev. gén. Bot. *40*: 328 (1928)
      A. gossypii (Ashby & Nowell) Guill. = *N. gossypii* Ashby & Nowell
   = *Spermophthora* Ashby & Nowell - Ann. Bot. *40*: 69 (1926)
      S. gossypii Ashby & Nowell
   1 or 2 species
   References: Pridham and Raper, 1950.

28. **Metschnikowia** Kamienski - Trav. Soc. imp. Nat. S. Pet. *30*: 363 (1899)
   M. bicuspidata (Metschnikoff) Kamienski
   ≡*Monospora* Metschnikoff, Virchows Arch. path. Anat. Physiol. *96*: 177 (1884), non Hochstetter, 1841
   ≡*Monosporella* Keilin - Parasitology *12*: 83 (1920)
   ≡*Metschnikowiella* Henkel, 1913, sensu Kudrjawzew, 1960
   = *Nadsoniomyces* Kudrjawzew, 1932 (? 1938) sensu Kudrjawzew, 1960
   = *Chlamydozyma* Wickerham - Mycologia *56*: 257 (1964)
      Ch. reukaufii Wickerh.
   6 species (Fig. 30d)
   References: Kudrjawzew, 1960; Miller, Barker and Pitt, 1967; Spencer, Phaff and Gardner, 1964; Fell and Hunter, 1968; Pitt and Miller, 1968; van Uden and Castelo-Blanco, 1961.

32. **Guilliermondella** Nadson & Krassilnikov - C. r. hebd. Séanc. Acad. Sci., Paris *187*: 307 (1928)
   G. selenospora Nadson & Krassilnikov (Fig. 31c)
   1 species
   References: von Arx, 1972.

Appendix

33. **Prototheca** Krüger - Hedwigia *33*: 263 (1894)
   P. zopfii Krüger
   6 species. The genus comprises colourless algae of the Protococcales, related to *Chlorella*
   References: Tubaki and Soneda, 1959; Cooke, 1968.

TORULOPSIDALES

1. Sexual forms forming chlamydospores or hyphae with clamps present ... 20
1. Sexual forms absent 2
2. Conidia borne on sterigmata present .... *Sterigmatomyces* (2)
2. Conidia borne on sterigmata absent 3
3. Septate hyphae and fission cells present 4
3. Fission cells absent, septate and pseudo-hyphae may be present 6
4. Budding cells present .... *Trichosporon* (13)
4. Budding cells absent 5
5. Fission cells borne in aerial hyphae .... cf. *Geotrichum*
5. Fission cells borne in prone hyphae, often containing one or more endospores .... *Protendomycopsis* (14)
6. Budding bipolar 7
6. Budding not bipolar 8
7. Cells lemon-shaped, fermentation .... *Kloeckera* (8)
7. Cells ellipsoid to cylindrical, no fermentation ..... *Schizoblastosporon* (11)
8. Cells often triangular, budding from angles .... *Trigonopsis* (6)
8. Cells not triangular 9
9. Cells often lunate or falcate .... *Selenotila* (7)
9. Cells not lunate or falcate 10
10. Strong acid production in sugar-containing media, cells frequently oval .....
   .... *Brettanomyces* (4)
10. Acid production in sugar-containing media weak or absent 11
11. Cells spherical, ovoid, or flask-shaped, with special nutritional requirement (fatty acids), budding unipolar on a more or less broad base ....
   .... *Pityrosporum* (5)
11. Cells spherical to cylindrical, no special requirement for fatty acids 12
12. No fermentation 13
12. Fermentation 15
13. Hyphae absent, inositol assimilated .... *Cryptococcus* (3)
13. Hyphae present or when absent inositol not assimilated 14
14. Carotenoid pigments present........................ *Rhodotorula* (9)
14. Carotenoid pigments absent 15
15. Hyphae present 16
15. Hyphae absent .... *Torulopsis* (1)

65

List of genera
(General reference: Lodder, 1970)

1. **Torulopsis** Berlese - Giorn. Vit. Enol. *3*: 52 (1895), Syll. Fung. *18*: 495 (1906)
   T. colliculosa (Hartm.) Sacc. (Fig. 31a)
   = *Asporomyces* Chaborski - Rech. Lévures thermoph. cryoph. p. 26 (1918)
      A. asporus Chaborski
   = *Paratorulopsis* Novák & Zsolt - Acta bot. hung. 7: 101 (1961) (pro maxima parte)
   about 40 species
   References: Lodder and Kreger-van Rij, 1952.

2. **Sterigmatomyces** Fell - Antonie van Leeuwenhoek *32*: 101 (1966)
   S. halophilus Fell (Fig. 31b)
   2 species.

3. **Cryptococcus** Kützing emend. Vuill. - Rev. gén. Sci. *12*: 741 (1901)
   C. neoformans (Sanfelice) Vuill. (Fig. 31c)
   = *Atelosaccharomyces* Beurm. & Gougerot - Tribune Méd. *42*: 502 (1909)
      A. busse-buschki Beurm. & Gougerot = *C. neoformans*
   = *Dioszegia* Zsolt - Bot. Közlem. 47: 63 (1957)
      D. hungarica Zsolt = *C. hungaricus* (Zsolt) Phaff & Fell
   12 species
   References: Lodder and Kreger-van Rij, 1952.

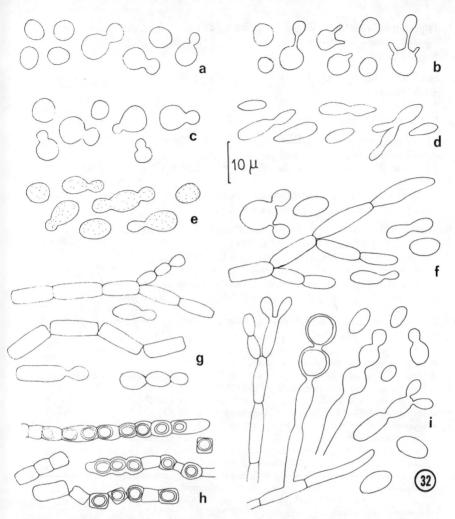

Fig. 32. a, *Torulopsis colliculosa*; b, *Sterigmatomyces halophilus*; c, *Cryptococcus neoformans*; d, *Brettanomyces bruxellensis*; e, *Rhodotorula glutinis*; f, *Candida tropicalis*, blastospores; g, *Trichosporon cutaneum*, blastospores and arthrospores; h, *Protendomycopsis dulcita*, arthrospores and endospores; i, *Syringospora (Candida) albicans*, hyphae, blastospores and chlamydospores (orig.).

4. **Brettanomyces** Kuff. & van Laer - Bull. Soc. chim. Belg. *30*: 270 (1921)
   B. bruxellensis Kuff. & van Laer (Fig. 31d)
   8 species
   References: Lodder and Kreger-van Rij, 1952.

5. **Pityrosporum** Sabour. - Malad. Cuir. chevelu II (1904)
   P. malassezii Sabour. = *P. ovale* (Bizz.) Cast. & Chalmers
   3 species
   References: Lodder and Kreger-van Rij, 1952.

6. **Trigonopsis** Schachner - Z. ges. Brauwesen *52*: 137 (1929)
    T. variabilis Schachner
1 species.

7. **Selenotila** Lagerh. - Ber. dt. bot. Ges. *10*: 531 (1892)
    S. nivalis Lagerh.
3 species
References: Yarrow, 1970.

8. **Kloeckera** Janke - Zentbl. Bakt. Parasitkde, Abt. 2, *59*: 310 (1923)
    Kloeckera apiculata (Reess) Janke
= *Pseudosaccharomyces* Klöcker - Zentbl. Bakt. Parasitkde, Abt. 2, *35*: 375 (1912)
    non Briosi & Farn
4 species, imperfect states of *Hanseniaspora*
References: Lodder and Kreger-van Rij, 1952.

9. **Rhodotorula** Harrison - Trans. R. Soc. Canada 3, *21*: 349 (1927)
    R. glutinis (Fres.) Harrison (Fig. 31e)
= *Eutorulopsis* Ciferri - Atti Ist. bot. Univ. Pavia 1, *2*: 147 (1925)
    E. dubia Cif. & Red. = *R. mucilaginosa* (Jörg.) Harrison
11 species. The genus is closely related to *Sporobolomyces*
References: Lodder and Kreger-van Rij, 1952; Hasegawa, 1965.

10. **Candida** Berkhout - Schimmelgesl. Monilia, Oidium etc. Utrecht, p. 72 (1923)
    Nom. gen. conservandum (Lanjouw & al., 1966, p. 241)
    C. vulgaris Berkhout = *C. tropicalis* (Cast.) Berkhout (Fig. 31f)
= *Anthomyces* Grüss - Ber. dt. bot. Ges. *35*: 746 (1917)
    A. reukaufii Grüss = *C. reukaufii* (Grüss) Diddens & Lodder (cf. *Metschnikowia*)
= *Apiotrichum* Stautz - Phytopath. Z. *3*: 163 (1931)
    A. porosum Stautz = *C. humicola* (Dasz.) Diddens & Lodder
= *Azymocandida* Novák & Zsolt - Acta bot. hung. 7: 100 (1961)
= *Azymoprocandida* Novák & Zsolt - l.c. p. 100 (1961)
= *Blastodendrion* (Ota) Cif. & Redaelli - Ann. mycol. 27: 243 (1929)
    B. krausii Ota = *C. guilliermondii* (Cast.) Lang. & Guerra (cf. *Pichia*)
= *Castellania* C. W. Dodge - Med. Mycology p. 246 (1935)
    C. bronchialis (Cast.) C. W. Dodge = *C. tropicalis* (Cast.) Berkhout
= *Endoblastoderma* Fischer & Brebeck - Morphol. Kahmpilze, Jena (1894)
    E. amycoides Fischer & Brebeck = *C. mycoderma* (Reess) Lodder & Kreger-van Rij
= *Microanthomyces* Grüss - Jb. wiss. Bot. *66*: 109 (1926)
    M. alpinus Grüss = *C. guilliermondii* (cf. *Pichia*)
= *Myceloblastanon* Ota - Jap. J. Dermatol. Urol. *28*: (1928)
    M. krausii Ota = *C. guilliermondii* (cf. *Pichia*)
= *Mycocandida* Lang. & Talice - Ann. Parasit. hum. comp. *10*: 56 (1932)
    M. mortifera (Red.) Lang. & Talice = *C. pseudotropicalis* (Cast.) Basgal
= *Mycokluyveria* Cif. & Red. - Mycopath. Mycol. appl. *4*: 54 (1947)
    M. vini (Desm.) Cif. & Red. = *C. mycoderma*
= *Nectaromyces* Syd. - Ann. mycol. *16*: 240 (1918)
    N. reukaufii (Grüss) Syd. = *C. reukaufii*
= *Parasaccharomyces* Beurm. & Gougerot - Trib. Méd. *42*: 502 (1909)
    P. sambergeri Beurm. & Gougerot
= *Parendomyces* Queyrat & Laroche - Bull. Mém. Soc. méd. Hôp. Paris, Sér. 3, *28*: 136 (1909)
    P. albus Queyrat & Laroche = *C. alba* (Queyrat & Laroche) Almeida
= *Pseudomonilia* Geiger - Zentbl. Bakt. Parasitenkde, Abt. 2, *27*: 97 (1910)
    P. albomarginata Geiger

about 90 species. The genus is a very unnatural one
References: Lodder, 1971.

1. Schizoblastosporon Ciferri - Arch. Protistenk. *71*: 405 (1930)
    Sch. starkeyi-henricii Ciferri
    1 species
    References: Roberts, 1960; di Menna, 1965.

2. Oosporidium Stautz - Phytopath. Z. *3*: 196 (1931)
    O. margaritiferum Stautz
    1 species.

3. Trichosporon Behrend - Berl. klin. Wschr. *27*: 464 (1890)
    T. cutaneum (Beurm. & al.) Ota (Fig. 31g)
    = *Geotrichoides* Lang. & Talice - Ann. Parasit. hum. comp. *10*: 67 (1932)
       G. cutaneus (Beurm. & al.) Lang. & Talice = *T. cutaneum*
    = *Proteomyces* Moses & Vianna - Mem. Inst. Oswaldo Cruz *5*: 192 (1913)
       P. infestans Moses & Vianna = *T. infestans* (Moses & Vianna) Cif. & Redaelli
    about 10 species
    References: Lodder and Kreger-van Rij, 1952.

4. Protendomycopsis Windisch - Beitr. Biol. Pfl. *41*: 340 (1965)
    P. domschii Windisch = *P. dulcita* (Berkh.) W. Gams & Domsch (Fig. 31h)
    The fungus is related to *Trichosporon cutaneum* (Beurm. & al.) Ota, but no
    blastospores could be found in the type culture and the cells of the hyphae or the
    arthrospores often contain 1-3 endospores. These were regarded as ascospores by
    the author of the genus.
    References: Gams and Domsch, 1969.

5. Syringospora Quinquand - Arch. Physiol. norm. path. *1*: 293 (1868)
    S. robinii Quinquand = *Candida albicans* (Robin) Berkhout = *Oidium albicans*
    Robin = *Syringospora albicans* (Robin) Dodge (Fig. 31i)
    = *Mycotorula* Will - Zentbl. Bakt. Parasitkde, Abt. 2, *46*: 226 (1916)
       M. craterica Will = *S. albicans*
    = *Mycotoruloides* Lang. & Talice - Ann. Parasit. hum. comp. *10*: 48 (1932)
       M. triadis Lang. & Talice = *S. albicans*
    = *Procandida* Novák & Zsolt - Acta bot. hung. *7*: 93 (1961)
       P. albicans (Robin) Novák & Zsolt = *S. albicans*
    1 species
    References: Van der Walt, 1969.

6. Rhodosporidium Banno - J. gen. appl. Microbiol. *13*: 192 (1967)
    R. toruloides Banno (perfect state of *Rhodotorula glutinis* (Fres.) Harrison)
    2 species.

7. Leucosporidium Fell & al. - Antonie van Leeuwenhoek *35*: 433 (1969)
    Leucosporidium scottii Fell & al.
    6 species.

8. Filobasidium Olive - J. Elisha Mitchell Sci. Soc. *84*: 261 (1968)
    F. floriforme Olive
    second species: *F. capsuligenum* (Fell & al.) Rodrigues de Miranda (Fig. 101e)
    References: Rodrigues de Miranda, 1972.

SPOROBOLOMYCETALES

1. Septate hyphae present, cultures never red or salmon 4

1. Septate hyphae absent or not predominant 2

2. Ballistospores spherical, carotenoids absent . . . . *Bullera* (2)

2. Ballistospores elongate, cultures often red or salmon by carotenoids 3

3. Thick-walled chlamydospores absent . . . . *Sporobolomyces* (1)

3. Sexual forms forming thick-walled chlamydospores present . *Aessosporon* (3)

4. Chlamydospores absent or not pigmented 5

4. Pigmented chlamydospores present 6

5. Hyphae without clamp connexions . . . . *Tilletiopsis* (4)

5. Hyphae with clamp connexions . . . . *Itersonilia* (5)

6. Chlamydospores glabrous . . . . *Sporidiobolus* (6)

6. Chlamydospores with projections, germinating with a 'phragmobasidium' forming ballistospores . . . . *Tilletiaria* (7)

List of genera

1. **Sporobolomyces** Kluyver & van Niel - Zentbl. Bakt. Parasitkde, Abt. 2, *63*: 19 (1924)
   S. roseus Kluyver & van Niel (Fig. 32)
   = *Amphierna* Grüss - Jb. wiss. Bot. *66*: 109 (1926)
      A. rubra Grüss = *S. roseus*
   14 species
   References: Lodder and Kreger-van Rij, 1952.

2. **Bullera** Derx - Ann. mycol. *28*: 11 (1930)
   B. alba (Hanna) Derx
   3 species
   References: Lodder and Kreger-van Rij, 1952.

3. **Aessosporon** van der Walt - Antonie van Leeuwenhoek *36*: 54 (1970)
   A. salmonicolor van der Walt, the sexual state of *Sporobolomyces salmonicolor* (Fischer & Brebeck) Kluyver & van Niel.

4. **Tilletiopsis** Derx - Ann. mycol. *28*: 3 (1930)
   T. minor Nyland
   2 or 3 species
   References: Nyland, 1949, 1950; Sowell and Korf, 1960; Moreau, 1963; Laffin and Cutter, 1959; Tubaki, 1952.

5. **Itersonilia** Derx - Bull. bot. Gard. Buitenzorg 3,*17*: 465 (1948)
   I. perplexans Derx
   3 species
   References: see *Itersonilia.*

6. **Sporidiobolus** Nyland - Mycologia *41*: 686 (1949)
   S. johnsonii Nyland
   1 species.

7. **Tilletiaria** Bandoni & Johri - Can. J. Bot. *50*: 39 (1972)
   T. anomala Bandoni & Johri
   1 species.

All the species of these orders that have been cultivated on laboratory media have developed either budding cells or slimy mycelium, which are insufficiently characteristic to distinguish between the genera. Identification is only possible by studying the structures developed on the living host plant. The following key may be useful for the recognition of these fungi:

1. Parasitic on Ericaceae or Theaceae, causing hyperplasia, hyphae absent in pure culture, budding cells cylindrical or filiform . . . . *Exobasidium* (3)

1. Not parasitic on Ericaceae 2

2. Parasitic on leaves of Quercus, Juglans, Carya and related trees, colonies white, budding cells ovoid or spherical . . . . *Microstroma* (4)

2. Not parasitic on Juglans or Quercus or colonies not white 3

3. Hyphae absent, colonies often salmon coloured or slightly brownish 4

3. Hyphae usually present, colonies whitish, cream coloured or greyish brown . . 5

4. Parasitic on Umbelliferae or Compositae . . . . *Protomyces* (2)

4. Parasitic on Rosaceae (Prunoideae), Betulaceae, Salicaceae, Quercus and other trees . . . . *Taphrina* (1)

5. Colonies spreading, hyphae quite broad; thick-walled chlamydospores often present, budding cells borne on hyphae . . . . cf. *Kabatiella*

5. Colonies mostly restricted, hyphae thin, hyaline 6

6. Budding cells present, ellipsoidal, fusiform or elongate, often in chains, parasitic on Gramineae, Caryophyllaceae, Polygonaceae, Compositae and other plants . . . . *Ustilago* (5)

6. Hyphae predominant, blastospores often absent or when present hypha-like, parasitic on monocotyledonous hosts . . . . *Tilletia* (6) *Urocystis* (7)
and other genera of the Ustilaginales

Cultures of *Kabatiella* species cannot usually be distinguished from those of *Aureobasidium (Pullularia) pullulans* (de Bary) Arn.

List of genera

1. **Taphrina** Fr. - Syst. mycol. *3*: 520 (1832)
   T. populina Fr.
= *Exoascus* Fuck. - Enum. Fungi Nassau p. 29 (1860)
   E. deformans (Berk.) Fuck. = *T. deformans* (Berk.) Tul.
= *Magnusiella* Sadebeck - Paras. Exoasc. *2*: 86 (1893)
   M. potentillae (Farl.) Sadebeck = *T. potentillae* (Farl.) Johans.
about 100 species (Fig. 33a)
References: Mix, 1949; Kramer, 1960.

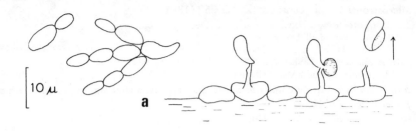

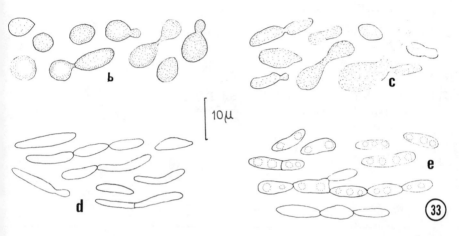

Fig. 33. a, *Sporobolomyces roseus*; b, *Taphrina deformans* (Berk.) Tul.; c, *Protomyces pachydermus* Thüm.; d, *Exobasidium vaccinii*; e, *Ustilago violacea* (Pers.) Roussel (orig.).

2. **Protomyces** Unger - Exanth. Pflanz. p. 341 (1833)
    P. macrosporus Unger
= *Protomycopsis* Magnus - Pilze Tirol p. 322 (1905)
    P. leucanthemi Magn.
= *Taphridium* Lagerh. & Juel - Bih. Svenska vet. Handl. *27*: 16 (1902)
    T. umbelliferarum (Rost.) Lagerh. & Juel = *P. umbelliferarum* Rost.
= *Volkartia* Maire - Bull. Soc. bot. Fr. *54*: 145 (1907)
    V. rhaetica (Volk.) Maire
about 15 species, all parasitic on Umbelliferae and Compositae, causing leaf galls
(Fig. 33b)
References: Tubaki, 1957.

3. **Exobasidium** Woronin - Verh. nat. Ges. Freiburg *4*: 397 (1867)
    E. vaccinii (Fuck.) Woronin (Fig. 33c)
about 10 species, all causing hyperplasia (galls) on Ericaceae and Theaceae
(Commelinaceae).- Pure cultures are known of 3 or 4 species
References: Graafland, 1953, 1960; Sundström, 1964.

4. **Microstroma** Niessl - Oest. bot. Z. *11*: 252 (1861)
   M. quercigenum Niessl = *M. album* (Desm.) Sacc.
   = *Helostroma* Pat. - Bull. Soc. mycol. Fr. *18*: 47 (1902)
      H. album (Desm.) Pat. = *M. album*
   = *Articulariella* Höhnel - Sber. Akad. Wiss. Wien *118*: 410 (1909)
      A. aurantiaca (Ellis & Mart.) Höhnel = *M. album*
   3 or 4 species, a quite common species is *M. juglandis* (Béreng.) Sacc. on leaves of
   Juglans regia and *Carya* species
   References: von Arx, 1957, 1963; Charles, 1935.

5. **Ustilago** (Pers.) Roussel - Fl. Calvados *2*: 47 (1806)
   U. segetum Pers. = *U. tritici* (Pers.) Jensen
   about 300 species, especially on Gramineae (Fig. 33d)
   References: Fischer, 1953; Zundel, 1953; Ainsworth and Sampson, 1950;
   Mundkur and Thirumalachar, 1952; Lindeberg, 1959.

6. **Tilletia** Tul. - Ann. Sci. nat., Bot., Sér. 3, 7: 112 (1847)
   T. caries (DC.) Tul.
   about 40 species, all on Gramineae
   References: see *Ustilago*.

7. **Urocystis** Rabenh. - Klotzsch Herb. mycol. *2*: 393 (1856)
   U. occulta (Wallr.) Rabenh.
   about 60 species
   References: see *Ustilago*.

PEZIZALES

Only species belonging to a few genera develop ascomata with ripe asci and ascospores in pure culture. Under natural conditions most of these grow on excrements. A suitable medium therefore is sterilized dung or oatmeal-agar enriched with dung of rabbits or other animals. Ascospores should be used for subculturing and they must be incubated in darkness at 37°C for the first two days and then for a further period in light at 15-19°C. Another good medium for the development of ascomata is soil-infusion agar. Species of *Pyronema* have to be cultivated on weak media containing K_3PO_4, KNO_3 and NH_4NO_3, 0.1% of each.

1. Ascomata minute, gymnocarpous, without excipulum, ascospores reticulate or spiny 2

1. Not combining above characters 3

2. Ascospores reticulate, dark, asci fasciculate . . . . *Ascodesmis* (1)

2. Ascospores spiny, hyaline, asci in a small number . . . . *Eleutherascus* (2)

3. Ascomata gymnocarpous, minute, often coalescing, orange, asci cylindrical, 8-spored, ascospores ellipsoidal, hyaline . . . . *Pyronema* (16)

3. Ascomata hemiangiocarpous or angiocarpous 4

4. Ascospores large, purplish when ripe, brown in age, thick-walled when young 5

4. Ascospores not thick-walled when young or remaining hyaline 6

5. Ascospores cohering in a cluster . . . . *Saccobolus* (4)

5. Ascospores not cohering in a cluster, singly ejaculated . . . . . . *Ascobolus* (3)

6. Ascospores thick-walled when young, non guttulate, hyaline 7

6. Ascospores thin-walled 8

7. Ascomata cylindrical or turbinate, ascospores with a sheath. . *Thecotheus* (6)

7. Ascomata discoid, carotenoids often present . . . . *Iodophanus* (5)

8. Ascus apex or ascus wall blue in iodine, ascomata discoid or cupulate, usually large . . . . *Peziza* (21)

8. Asci not blue in iodine 9

9. Ascomata at least in younger states angiocarpous, spherical or turbinate . . 10

9. Ascomata hemiangiocarpous, usually discoid 14

10. Colonies spreading quickly, expanding hyphae broad, ascomata spherical or turbinate, at base with thick-walled appendage-like hyphae, asci cylindrical, 8-spored, ascospores nearly spherical . . . . *Orbicula* (15)

10. Not combining above characters 11

11. Ascomata with bright, often coiled appendages, asci clavate . . . . *Lasiobolidium* (13)

11. Ascomata without such appendages 12

12. Ascomata small, thin-walled, asci ovoidal, 8-spored ...*Cleistothelebolus* (12)

12. Not combining above characters 13

13. Ascomata with a dark wall, asci 8-spored, cylindrical .... *Warcupia* (14)

13. Ascomata with a bright wall, asci 8- or many-spored .... *Thelebolus* (7)

14. Ascospores not guttulate, carotenoids usually absent (usually coprophilous) .
.... 15

14. Ascospores guttulate, carotenoids usually present 18

15. Asci opening by an operculum 16

15. Asci opening by a slit or a tear 17

16. Ascomata setose .... *Lasiobolus* (9)

16. Ascomata not setose .... *Coprotus* (11)

17. Ascomata setose .... *Trichobolus* (10)

17. Ascomata not setose .... *Ascozonus* (8)

18. Ascomata glabrous, carotenoids present, ascospores not guttulate 19

18. Ascomata hairy or setose, ascospores guttulate, carotenoids absent. 20

19. Ascospores spherical .... *Lamprospora* (17)

19. Ascospores ellipsoidal or fusiform.................... *Octospora* (20)

20. Ascospores spherical .... *Sphaerosporella* (18)

20. Ascospores ellipsoidal or fusiform.................... *Trichophaea* (19)
(for other genera, not known in pure culture compare Rifai, 1968; Eckblad, 1968; Korf, 1973)

Species of many genera of the Pezizales produce mycelium but no ascomata in pure culture. Some species develop conidial states belonging to genera such as *Dichobotrys, Chromelosporium* or *Oedocephalum.*

List of genera

1. **Ascodesmis** van Tieghem - Bull. Soc. bot. Fr. *23*: 275 (1876)
   A. nigricans van Tieghem
   5 species (Fig. 34)
   References: Obrist, 1961.

2. **Eleutherascus** v. Arx - Persoonia *6*: 377 (1971)
   E. lectardii (Nicot) v. Arx

3. **Ascobolus** Pers. ex Fr. - Syst. mycol. *2*: 161 (1822)
   A. pezizoides Pers. apud Gmel. = *A. furfuraceus* Pers. ex Fr.
   = *Crouaniella* (Sacc.) Lamb. - Fl. mycol. Belg., suppl. *1*: 320 (1887)
     C. murina (Fuckel) Lamb. = *A. brassicae* Crouan

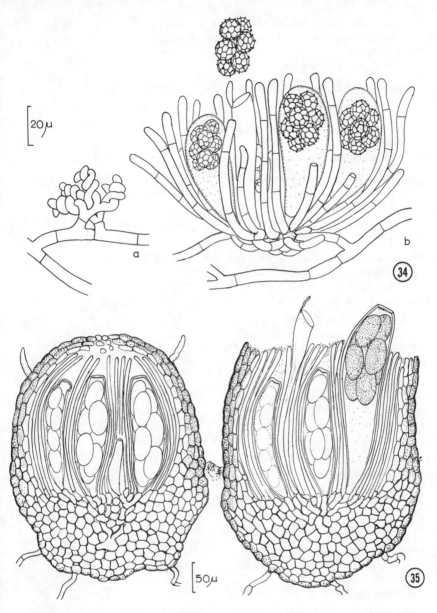

Fig. 34. *Ascodesmis sphaerospora* Obrist, a, ascogonial coils; b, mature ascoma (orig.).
Fig. 35. *Ascobolus immersus*, two ascomata, closed or exposing the hymenium (orig.).

= *Dasyobolus* Sacc. - Syll. Fung. *11*: 421 (1895)
  D. immersus (Pers.) Sacc. = *A. immersus* Pers.
= *Seliniella* v. Arx & Müller - Acta bot. neerl. *4*: 118 (1955)
  S. macrospora v. Arx & Müll. = *A. immersus*

= *Sphæridiobolus* Boud. - Bull. Soc. mycol. Fr. *1*: 108 (1885)
    S. hyperboreus (Karst.) Boud. = *A. hyperboreus* Karst.
about 50 species (Fig. 35)
References: van Brummelen, 1967.

4. **Saccobolus** Boud. - Ann. Sci. nat., Bot., Sér. 5, *10*: 228 (1869)
    S. kervernii (Crouan) Boud.
= *Ornithascus* Vel. - Monogr. Discom. Bohem. *1*: 368 (1934)
    O. corvinus Vel. = *S. dilutellus* (Fuckel) Sacc.
18 species
References: van Brummelen, 1967.

5. **Iodophanus** Korf apud Kimbrough & Korf - Am. J. Bot. *54*: 18 (1967)
    I. carneus (Pers.) Korf (= *Ascophanus carneus* (Pers.) Boud.) (Fig. 36a)
10 species
References: Kimbrough and Korf, 1967; Kimbrough, Luck-Allen and Cain, 1969; Korf, 1958; Rifai, 1968.

6. **Thecotheus** Boud. - Ann. Sci. nat., Bot., Sér. 5, *10*: 235 (1869)
    T. pelletieri (Crouan & Crouan) Boud.
second species: *T. cinereus* (Crouan & Crouan) Chenantais
References: Kimbrough, 1966, 1969; Kimbrough and Korf, 1967.

7. **Thelebolus** Tode ex Fr. - Syst. mycol. *2*: 306 (1823)
    Thelebolus stercoreus Tode ex Fr.
= *Ryparobius* Boud. - Ann. Sci. nat., Bot., Sér. 5, *10*: 237 (1869)
    R. cookei Boud. = *T. crustaceus* (Fuck.) Kimbrough
= *Pezizula* Karst. - Bidr. Kenn. Finl. Nat. Folk *1*: 9 (1871)
    P. conformis Karst.
= *Chilonectria* Sacc. - Michelia *1*: 279 (1878)
    C. cucurbitula (Curr.) Sacc.
several species, delimitation confusing (Fig. 36)
References: Kimbrough and Korf, 1967; Kimbrough, 1966.

8. **Ascozonus** (Renny) Boud. - Bull. Soc. mycol. Fr. *1*: 109 (1885)
    A. cunicularius (Boud.) Boud.
= *Streptotheca* Vuill. - J. Bot., Paris *1*: 33 (1887)
    S. boudieri Vuill. = *A. crouani* (Renny) Boud.
3 species
References: Kimbrough and Korf, 1967.

9. **Lasiobolus** Sacc. - Bot. Zentbl. 2, *18*: 220 (1884)
    L. equinus (Müll. ex Gray) Karst.
4 species
References: Kimbrough and Korf, 1967.

10. **Trichobolus** (Sacc.) Kimbr. & Cain apud Kimbr. & Korf - Am. J. Bot. *54*: 20 (1967)
    T. zukalii (Heimerl) Kimbr.
3 species
References: Kimbrough and Korf, 1967; Kimbrough, 1966.

11. **Coprotus** Korf & Kimbr. - Amer. J. Bot. *54*: 21 (1967)
    C. sexdecimsporus (Crouan & Crouan) Kimbr. & Korf

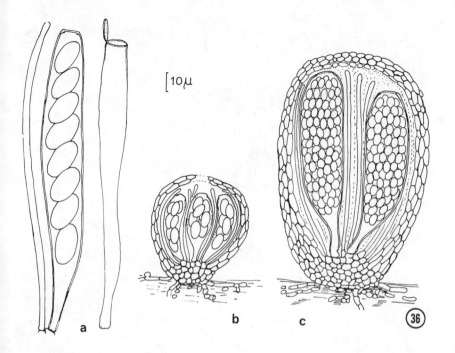

Fig. 36. a, *Iodophanus carneus*, asci and paraphysis; b, *Thelebolus microsporus* (Berk. & Br.) Kimbrough: c. *Thelebolus crustaceus* (Fuck.) Kimbrough, ascomata (orig.).

7 or more species.

12. **Cleistothelebolus** Malloch & Cain - Can. J. Bot. *49*: 851 (1971)
    C. nipigonensis Malloch & Cain
    1 species.

13. **Lasiobolidium** Malloch & Cain - Can. J. Bot. *49*: 853 (1971)
    L. spirale Malloch & Cain
    2 species.

14. **Warcupia** Paden & Cameron - Can. J. Bot. *50*: 999 (1972)
    W. terrestris Paden & Cameron
    1 species.

15. **Orbicula** Cooke - Handb. Brit. Fungi *2*: 296 (1871)
    O. parietina (Schrader ex Fr.) Hughes
    1 species
    References: Hughes, 1951.

16. **Pyronema** Carus - Nova Acta Leop. 1, *17*: 370 (1835)
    P. marianum Carus = *P. omphalodes* (Bull. ex St.-Amans) Fuckel
    second species: *P. domesticum* (Sow. ex Gray) Sacc. (Fig. 37)
    References: Moore and Korf, 1963; Moore, 1962, 1963; Robinson, 1926.

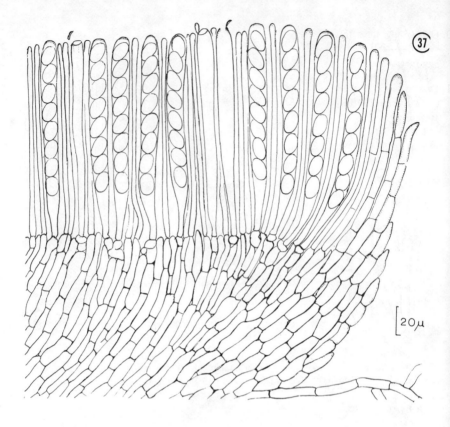

Fig. 37. *Pyronema domesticum*, part of an apothecium with asci and paraphyses (orig.).

17. **Lamprospora** de Not. - Commentario Soc. critt. ital. *1*: 388 (1864)
    L. miniata de Not.
    synonyms see Rifai, 1968
    7 species.

18. **Sphaerosporella** (Svrček) Svrček & Kubička - Česká Mykol. *15*: 66 (1961)
    S. brunnea (Alb. & Schw. ex Fr.) Svrček & Kubička
    = *Sphaerospora* auct. non Sacc.
    conidial state: *Dichobotrys*
    2 or 3 species
    References: Rifai, 1968; Cain and Hastings, 1956.

19. **Trichophaea** Boudier - Bull. Soc. mycol. Fr. *1*: 105 (1885)
    T. woolhopeia (Cooke & Phill.) Boudier
    12 or more species
    conidial state: *Dichobotrys*
    References: Kanouse, 1958.

0. **Octospora** Hedw. ex S.F. Gray - Nat. Arrang. Br. Pl. *1*: 666 (1821)
   O. leucoloma Hedw. ex S.F. Gray
synonyms see Rifai, 1968; Korf, 1973.

1. **Peziza** L. ex St-Amans emend. Boudier - Bull. Soc. mycol. Fr. *1*: 103 (1885)
   P. vesiculosa Bull. ex St-Amans
References: Rifai, 1968; Eckblad, 1968; Korf, 1973.

Only a few species, mainly those belonging to the genera *Sclerotinia, Monilinia, Gloeotinia, Rutstroemia, Tympanis* or *Pezicula*, are known to develop ascomata in pure culture and special conditions are required. Other species develop only a sterile mycelium or conidial states belonging to the Melanconiales, Sphaeropsidales or Moniliales. Well known imperfect genera of Helotiales are *Botrytis, Monilia, Cryptosporiopsis, Sporonema, Gloeosporidiella, Marssonina, Entomosporium, Phloeosporella, Pleurophomella, Micropera, Hainesia, Ceuthospora* or *Leptostroma*. A large number of described species have not yet been cultivated and cultivation experiments were often unsuccessful (e.g. with *Rhytisma* species).

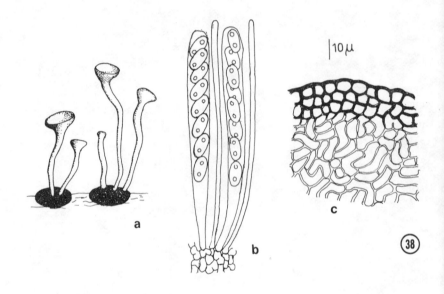

Fig. 38. *Sclerotinia sclerotiorum*, a, sclerotia with apothecia; b, asci and paraphyses; c, part of a sclerotium (orig.).

List of genera,
known to develop conidial states or ascomata in pure culture
(for keys compare Dennis, 1968; Korf, 1973)

1. **Sclerotinia** Fuckel - Symb. Mycol. p. 330 (1870)
   S. libertiana Fuckel = *S. sclerotiorum* (Lib.) de Bary (Fig. 38)
   = *Botryotinia* Whetzel - Mycologia 37: 679 (1941)
   B. fuckeliana (de Bary) Whetzel = *S. fuckeliana* (de Bary) Fuckel

= *Ciborinia* Whetzel - Mycologia *37*: 668 (1945)
   C. candolleana (Lév.) Whetzel = *S. candolleana* (Lév.) Fuckel
= *Myriosclerotinia* Buchwald - Friesia *3*: 291 (1947)
   M. scirpicola (Rehm) Buchwald = *S. scirpicola* Rehm
= *Septotinia* Whetzel - Mycologia *29*: 128 (1937)
   S. podophyllina Whetzel = *Sclerotinia podophyllina* (Whetzel) v. Arx
= *Stromatinia* Boud. - Hist. Classif. Discom. Europ. p. 108 (1907)
   S. rapulum (Bull. ex Fr.) Boud. = *S. rapula* (Bull. ex Fr.) Rehm
= *Whetzelinia* Korf & Dumont - Mycologia *64*: 250 (1972)
   W. sclerotiorum (Lib.) Korf & Dumont = *S. sclerotiorum*
conidial states: *Botrytis, Septotis, Myrioconium* (spermatia, borne on phialides)
about 40 species, mostly parasitic on higher plants
References: Buchwald, 1949; Dennis, 1956, 1968.

2. **Monilinia** Honey - Mycologia *20*: 153 (1928)
   M. fructicola (Wint.) Honey (= *Sclerotinia fructicola* (Wint.) Rehm)
conidial state: *Monilia*
about 10 species
References: Dennis, 1956; Whetzel, 1945; Honey, 1936.

3. **Gloeotinia** Wilson & al. - Trans. Br. mycol. Soc. *37*: 31 (1954)
   G. temulenta (Prill. & Delacr.) Wilson & al.
1 species, on caryopses of grasses and cereals
References: Dennis, 1956.

4. **Rutstroemia** Karst. - Mycol. Fenn. *1*: 12 (1871)
   R. firma (Pers. ex Fr.) Karst.
about 20 species, mostly saprophytic
References: White, 1941; Dennis, 1956.

5. **Dermea** Fr. - Syst. Orb. Veg. p. 114 (1825)
   D. cerasi (Pers. ex Fr.) Fr.
conidial state: *Micropera (Gelatinosporium, Chondropodium, Micula, Brunchorstia)*
about 20 species
References: Groves, 1946.

6. **Discohainesia** Nannf. - Nova Acta Reg. Soc. Sci. Upsal. IV, *8*,2: 88 (1932)
   D. oenotherae (Cooke & Ellis) Nannf.
conidial state: *Hainesia*
1 species.

7. **Tympanis** Tode ex Fr. - Syst. Mycol. *2*: 172 (1822)
   T. saligna Tode ex Fr.
conidial state: *Pleurophomella*
about 40 species
References: Groves, 1952.

8. **Pezicula** Tul. - Sel. Fung. Carpol. *3*: 182 (1865)
   P. carpinea (Pers. ex Fr.) Tul.
= *Dermatina* (Sacc.) Höhnel - Sber. Akad. Wiss. Wien *118*: 1521 (1909)
   D. fagi (Phill.) Höhnel
= *Neofabraea* Jackson - Rep. Oreg. Exp. Stn p. 187 (1911/12)
   N. malicorticis Jackson = *P. malicorticis* (Jackson) Nannf.

= *Phaeangium* Sacc. - Syll. Fung. *16*: 764 (1901)
    P. rubi (Bäuml.) Sacc. = *Pezicula rubi* (Lib.) Niessl
conidial state: *Cryptosporiopsis*
about 25 species. Common on conifers is *P. livida* (Berk. & Br.) Rehm
References: Johansen, 1949.

9. **Ocellaria** (Tul.) Karst. - Mycol. Fenn. *1*: 21 (1871)
    O. ocellata (Pers.) Schroet.
conidial state: *Cryptosporiopsis scutellata* (Otth) Petrak
1 species on *Salix* and *Populus*.

10. **Crumenulopsis** Groves - Can. J. Bot. *47*: 48 (1969)
    C. pinicola (Fr.) Groves (= *Crumenula pinicola* (Fr.) Karst.)
second species: *Crumenulopsis sororia* (Karst.) Groves with *Digitosporium piniphilum* Gremmen as conidial state.

11. **Godronia** Moug. & Lév. in Moug. - Consid. gén. vég. spont. Dép. Vosges p. 355
    (1846)
    G. muehlenbeckii Moug. & Lév.
= *Scleroderris* (Pers. ex Fr.) Bon. - Handb. allg. Mykol. p. 201 (1851)
    S. ribis (Fr.) Thüm. = *G. ribis* (Fr.) Seaver
= *Crumenula* de Not. - Comm. Soc. crittog. ital. *1*: 365 (1864)
    C. urceolata (Schmidt ex Fr.) de Not. = *G. urceolata* (Schmidt ex Fr.) Karst.
conidial states: *Topospora* Fr., *Chondropodiella* Höhnel
about 25 species
References: Groves, 1965; Schläpfer-Bernhard, 1968; Eriksson, 1970.

12. **Ascocalyx** Naumov - Bolezni rast. *14*: 138 (1925)
    A. abietis Naumov
= *Gremmeniella* Morelet - Bull. Soc. Sci. nat. Arch. Toulon Var *183*: 9 (1969)
    G. abietina (Lagerb.) Morelet = *Crumenula abietina* Lagerb. = *A. abietina* (Lagerb.)
    Schläpfer
= *Lagerbergia* Reid ap. Dennis - Kew Bull. *25*: 350 (1971)
    L. abietina (Lagerb.) Reid = *A. abietina*
conidial states: *Brunchorstia, Bothrodiscus*
3 species on conifers
References: Schläpfer-Bernhard, 1968.

13. **Leptotrochila** Karst. - Mycol. Fenn. *1*: 22 (1871)
    L. radians (Rob.) Karst.
= *Fabraea* Sacc. - Michelia *2*: 331 (1882)
    F. ranunculi (Fr.) Karst = *L. ranunculi* (Fr.) Schüepp
= *Ephelina* Sacc. - Syll. Fung. *8*: 585 (1889)
    E. lugubris (de Not.) Höhnel = *L. lugubris* (de Not.) Schüepp
= *Placopeziza* Höhnel - Sber. Akad. Wiss. Wien *125*: 52 (1916)
    P. phyteumatis (Fuckel) Höhnel = *L. phyteumatis* (Fuckel) Schüepp
conidial state: *Sporonema*
15 species, parasitic on plants
References: Schüepp, 1959

14. **Drepanopeziza** (Kleb.) Höhnel - Ann. mycol. *15*: 332 (1917)
    D. ribis (Kleb.) Höhnel
conidial states: *Gloeosporidiella, Marssonina*

84

8 species, parasitic on *Ribes, Populus, Salix*, etc.
References: Rimpau, 1961; Gremmen, 1965.

6. **Diplocarpon** Wolf - Bot. Gaz. *54*: 231 (1912)
   D. rosae Wolf
   = *Entomopeziza* Kleb. - Vortr. Gesamtgeb. Bot. *1*: 33 (1914)
   E. soraueri Kleb. = *D. soraueri* (Kleb.) Nannf. = *D. maculatum* (Atk.) Jørst.
   conidial states: *Actinonema, Marssonina, Entomosporium*
   4 species, parasitic on Rosaceae.

5. **Blumeriella** v. Arx - Phytopath. Z. *42*: 164 (1961)
   B. jaapii (Rehm) v. Arx
   = *Higginsia* Nannf. - Nova Acta Reg. Soc. Sci. Upsal. 4, *8*,2: 173 (1932) non *Higginsia* Pers. (1805)
   H. hiemalis (Higgins) Nannf. = *B. jaapii*
   conidial state: *Phloeosporella*
   1 or 2 species, parasitic on *Prunus*.

7. **Pyrenopeziza** Fuckel - Symb. Mycol. p. 293 (1870)
   P. chailletii Fuckel
   synonyms see Hütter, 1958
   conidial state: *Phialophora*
   about 40 species are described as herbarium specimens
   References: Gremmen, 1954, 1955, 1956; Hütter, 1958.

8. **Mollisia** (Fr.) Karst. - Mycol. Fennica *1*: 15 (1871)
   M. cinerea (Batsch ex Mérat) Karst.
   conidial state: *Phialophora*
   about 60 species
   References: Le Gal and Mangenot, 1958-1966; Dennis, 1968.

9. **Coryne** Tul. - Sel. Fung. Carpol. *3*: 190 (1865)
   C. sarcoides (Jacquin ex S. F. Gray) Tul.
   = *Ascocoryne* Groves & Wilson - Taxon *16*: 35 (1967)
   conidial state: *Phialophora*-like, but often forming sporodochia and described as
   Pirobasidium sarcoides Höhnel
   2 or 3 species, common on stumps and fallen logs. Characteristic is the formation
   of a brownish red pigment in vivo and in vitro.

EUROTIALES

1. Ascomata submerged, spherical or obcampanulate when erumpent, surrounded by an amorphous mass, asci fasciculate, broadly clavate, ascospores
ellipsoidal; conidia in acropetal chains present (*Hormoconis resinæ*) . . . .
. . . . *Amorphotheca* (54)

1. Not combining above characters 2

2. Ascomata are the enlarged ascogonial cells, cyst-like (isolated from bees or
honey) 3

2. Ascomata not cyst-like 4

3. Ascomata small, about 30μm in diameter, ascospores spherical, not in balls. .
. . . .*Bettsia* (53)

3. Ascomata larger, ascospores not spherical, in balls (many-spored asci) . . . . . .
. . . . *Ascosphaera* (52)

4. Ascomata stalked, small, with a thin, crust-like wall, asci 2-8-spored, evanescent, ascospores ellipsoidal; conidia with a truncate base borne in basipetal
succession often present . . . . *Monascus* (51)

4. Not combining above characters 5

5. Ascomata hemispherical, with a broad base, large, often aggregated in a
crust, reddish, ascospores ellipsoidal, ornamented 6

5. Not combining above characters 7

6. Thermophilic, asci not catenulate . . . .*Thermoascus* (14)

6. Not thermophilic, growing at temperatures lower than 35°, asci catenulate . .
. . . .*Aphanoascus* (15)

7. Conidia borne in true chains on phialides present. 8

7. Conidia borne in chains on phialides absent 20

8. Wall of the ascomata composed of a network of pigmented hyphae ending in
appendage-like structures, asci catenulate, ascospores ellipsoidal, conidial
state is *Acremonium* . . . .*Sagenoma* (13)

8. Not combining above characters 9

9. Wall of the ascomata surrounded by thick-walled 'hüllecells', ascospores
often reddish or yellow, conidial state is *Aspergillus* (biseriate) 10

9. Hüllecells absent 11

10. Initials filamentous, usually coiled, ascomata usually discrete. *Emericella* (3)

10. Initial is a large, subglobose cell, ascomata clustered . . . . *Fennellia* (2)

11. Wall of the ascomata thin, composed of flattened cells or hyphae, or absent
. . . . 12

11. Wall of the ascomata at least partly pseudoparenchymatous, often sclero-
tium-like or stromatic 18

12. Wall composed of a loose network of hyphae, asci clavate, ascospores ellipsoidal, conidial state is *Aspergillus* (uniseriate) . . . . *Warcupiella* (4)

12. Wall composed of hyphal elements or flattened cells or absent, asci mostly spherical 13

13. Wall of the ascomata composed of interwoven hyphae or absent, conidial state is *Penicillium* or *Paecilomyces* 14

13. Wall of the ascomata composed of flattened cells or hyphae or absent, conidial state is *Aspergillus* (mostly uniseriate) 16

14. Asci catenulate (in short chains) . . . . *Talaromyces* (10)

14. Asci not catenulate, borne from croziers 15

15. Ascomata with a hyphal wall, initials are single coiled hyphae, conidial state is *Penicillium* . . . . *Hamigera* (11)

15. Ascomata without wall, ascogonia coiled around a swollen antheridium, conidial state is *Paecilomyces* . . . . *Byssochlamys* (1)

16. Ascomata tomentose, conidia clavate or ovate; osmophilic *Chaetosartorya* (6)

16. Ascomata not tomentose, conidia mostly roundish in outline 17

17. Wall of the ascomata composed of a single layer of flattened cells or absent, conidia 4-10 μm in size; osmophilic . . . . *Eurotium* (5)

17. Wall of the ascomata composed of irregularly flattened cells or hyphae, conidia 2-3 μm in size; not osmophilic . . . . *Neosartorya* (7)

18. Ascomata sclerotioid, conidial state is *Penicillium* . . . . *Eupenicillium* (12)

18. Ascomata with a pseudoparenchymatous wall or immersed in a stroma, conidial state is *Aspergillus* 19

19. Ascomata immersed in an elongated, dark stroma, *Aspergillus*-state biseriate . . . . *Syncleistostroma* (9)

19. Ascomata single, spherical, with a pseudoparenchymatous or sclerotioid wall, *Aspergillus*-state uniseriate, conidia 5-8 μm long, ellipsoidal or clavate . . . . *Hemicarpenteles* (8)

20. Ascomata with a peridium composed of interwoven hyphae, often with appendages or without any peridium (*Gymnoascaceae*) 21

20. Ascomata with a compact wall composed of angular or flattened cells or of dense layers of filaments 39

21. Ascospores spherical, 10-13 μm in diameter, spiny, asci borne in pairs or in small numbers on a coil, peridium absent . . . . cf. *Eleutherascus*

21. Ascospores smaller than 10 μm 22

22. Ascospores ellipsoidal or fusiform 23

22. Ascospores spherical, oblate or lenticular 25

23. Ascospores striate by flutes, ascomata without distinct peridial hyphae . . . . *Byssoascus* (35)

48. Ascospores with a rim, smooth, *Chrysosporium-* *(Trichophyton-)*conidia present (keratinophilic) .... *Keratinophyton* (45)

48. Ascospores finely echinulate, conidia absent (not keratinophilic) .... *Xanthothecium* (46)

49. Ascomata stipitate, bright, ascospores ellipsoidal, arthric conidia present (keratinophilic) .... *Onygena* (18)

49. Not combining above characters 50

50. Ascospores ornamented or with appendages 51

50. Ascospores smooth 54

51. Ascospores with winged appendages, conidial state is *Acremonium* .... *Emericellopsis* (43)

51. Ascospores without winged appendages 52

52. Ascospores larger than 10 μm, spherical .... cf. *Roumegueriella*

52. Ascospores smaller or not spherical 53

53. Ascomata dark, ascospores brown, reticulate .... *Hapsidospora* (41)

53. Ascomata red or brownish, ascospores light, reticulate ..... *Anixiopsis* (44)

54. Ascomata with a wall composed of plates of radiating cells, with sutures (cephalothecoid) 55

54. Ascomata with a continuous wall composed of isodiametrical or flattened cells or of hyphal filaments 56

55. Ascospores short cylindrical .... *Cryptendoxyla* (40)

55. Ascospores reniform .... *Fragosphaeria* (39)

56. Ascomata thick-walled, aggregated to a crust, ascospores ellipsoidal, brown, *Verticillium*-conidial state present .... *Ephemeroascus* (55)

56. Not combining above characters 57

57. Ascospores small, hyaline or brown, often spherical or nearly so 58

57. Ascospores longer than 6 μm, ellipsoidal, hyaline .... see Pezizales

58. Ascomata smooth or nearly so 59

58. Ascomata with appendages 61

59. Ascomata small, with a brown wall composed of hyphal filaments, ascospores nearly spherical, hyaline, arthroconidia present ...... *Xylogone* (49)

59. Ascomata with a wall composed of angular or flattened cells........... 60

60. Ascomata with a thin wall composed of flattened cells, ascospores brownish, blastoconidia or phialoconidia present .... *Pseudeurotium* (48)

60. Ascomata with a thick, pseudoparenchymatous wall, ascospores hyaline, spherical, often with lateral bubbles................ *Nigrosabulum* (42)

61. Appendages of the ascomata bright, spirally coiled, ascospores spherical, often with lateral bubbles, nearly hyaline .... *Pleuroascus* (50)

61. Appendages usually long, pigmented, ascospores brownish . . . .
. . . . *Arachnomyces* (47)

List of genera

1. **Byssochlamys** Westling - Svensk. Bot. Tidskr. *3*: 134 (1909)
   B. nivea Westling (Fig. 40)
   conidial state: *Paecilomyces*
   second species: *B. fulva* Olliver & Smith
   References: Brown and Smith, 1957; Stolk and Samson, 1971.

2. **Fennellia** Wiley & Simmons - Mycologia *65*: 936 (1973)
   F. flavipes Wiley & Simmons = *Aspergillus flavipes* (Bainier & Sartory) Thom & Church.

3. **Emericella** Berk. & Br. in Berk. - Introd. crypt. Bot. p. 340 (1857)
   E. variecolor Berk. & Br.
   = *Diplostephanus* Langeron - C. r. Paris *87*: 344 (1922)
   D. nidulans (Eidam) Langeron = *E. nidulans* (Eidam) Vuill.
   = *Inzengaea* Borzi - Jb. wiss. Bot. *16*: 450 (1884)
   I. erythrospora Borzi = *E. variecolor*
   conidial state: *Aspergillus, Aspergillus nidulans* group
   about 12 species
   References: Benjamin, 1955; Raper and Fennell, 1965.

4. **Warcupiella** Subram. - Curr. Sci. *41*: 757 (1972, November)
   W. spinulosa (Warcup) Subram. = *Aspergillus spinulosus* Warcup in Raper & Fennell (1965)
   ≡ *Sporophormus* Malloch & Cain - Can. J. Bot. *50*: 2624 (1972, December).

5. **Eurotium** Link ex Fr. - Syst. mycol. *3*: 331 (1829)
   E. herbariorum (Pers.) Link ex Fr.
   = *Edyuillia* Subram. - Curr. Sci. *41*: 756 (1972)
   E. athecia (Raper & Fennell) Subram. = *Aspergillus athecius* Raper & Fennell (The genus Aspergillus p. 183, 1965) = *Eurotium athecium* (Raper & Fennell) v. Arx *comb. nov.*
   = *Gymneurotium* Malloch & Cain - Can. J. Bot. *50*: 2619 (1972)
   G. athecium (Raper & Fennell) Malloch & Cain = *E. athecium*
   conidial state: *Aspergillus, Aspergillus glaucus* group
   about 20 species (Fig. 42)
   References: Benjamin, 1955; Raper and Fennell, 1965; Hadlok and Stolk, 1969.

6. **Chaetosartorya** Subram. - Curr. Sci. *41*: 761 (1972, November)
   Ch. chrysellus (Kwon & Fennell) Subram. (= *Aspergillus chrysellus* Kwon & Fennell)
   ≡ *Harpezomyces* Malloch & Cain - Can. J. Bot. *50*: 2624 (1972, December)
   second species: *Ch. cremea* (Kwon & Fennell) Subram. = *Aspergillus cremeus*
   References: Raper and Fennell, 1965.

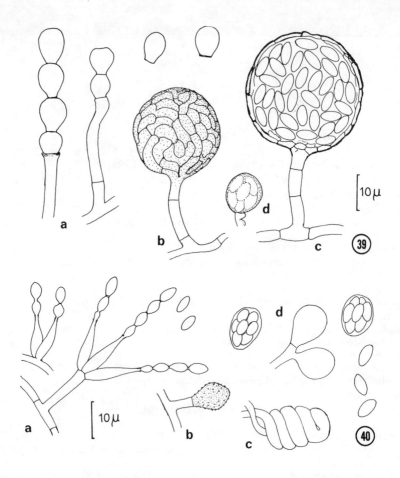

Fig. 39. *Monascus ruber* van Tieghem, a, conidial state; b, c, ascomata; d, ascus (orig.).
Fig. 40. *Byssochlamys nivea*, a, conidial state (*Paecilomyces*); b, chlamydospore; c, ascogonial coil; d, asci and ascospores (orig.).

7. **Neosartorya** Malloch & Cain - Can. J. Bot. *50*: 2620 (1972)
     N. fischeri (Wehmer) Malloch & Cain (Fig. 41)
   = *Sartorya* sensu Benjamin (1955) and Subramanian (1972)
   conidial state: *Aspergillus, Aspergillus fumigatus* group
   7 species
   References: Raper and Fennell, 1965.

8. **Hemicarpenteles** Sarbhoy & Elphick - Trans. Br. mycol. Soc. *51*: 156 (1968)
     H. paradoxus Sarbhoy & Elphick (= *Aspergillus acanthosporus* Udagawa & Takada)
   ≈ *Sclerocleista* Subram. - Curr. Sci. *41*: 757 (1972)
     S. ornata (Raper & al.) Subram. = *Aspergillus ornatus* Raper & al. (Mycologia *45*: 678,

92

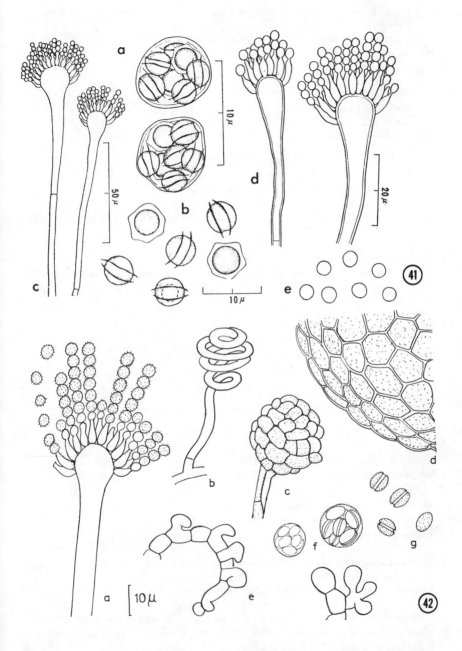

Fig. 41. *Neosartorya fischeri* (Wehmer) var. *glabra* (Fennell & Raper) Malloch & Cain, a, asci, b, ascospores, c, d, conidial heads (*Aspergillus*); e, conidia (from Udagawa and Kawasaki, 1968).

Fig. 42. *Eurotium amstelodami* Mangin, a, *Aspergillus* conidial state; b, ascogonial coil; c, young ascoma; d, part of the wall of an ascoma; e, ascogenous hyphae; f, asci; g, ascospores (orig.).

1953) =*Neosartoria ornata* (Raper & al.) Malloch & Cain = *Hemicarpenteles ornata* (Raper & al.) v. Arx *comb. nov.*

other species: *Hemicarpenteles thaxteri* (Subram.) v. Arx *comb. nov.* = *Sclerocleista thaxteri* Subram. (Curr. Sci. *41*: 757, 1972) = *Aspergillus citrisporus* Höhnel.

9. **Syncleistostroma** Subram. - Curr. Sci. *41*: 756 (1972, November)
   S. alliacea (Thom & Church) Subram. = *Aspergillus alliaceus* Thom & Church
   (see Raper and Fennell, 1965)
   ≡ *Petromyces* Malloch & Cain - Can. J. Bot. *50*: 2623 (1972, December).

10. **Talaromyces** Benjamin - Mycologia *47*: 681 (1955)
    T. vermiculatus (Dang.) Benjamin = *T. flavus* (Klöcker) Stolk & Samson
    conidial state: *Penicillium, Paecilomyces*
    about 16 species (Fig. 43)
    References: Udagawa, 1966; Udagawa and Takada, 1967; Stolk and Samson, 1972.

11. **Hamigera** Stolk & Samson — Persoonia *6*: 342 (1971)
    H. avellanea Stolk & Samson (= *Penicillium avellaneum* Thom & Turesson)
    second species: *Hamigera striata* Stolk & Samson (= *Penicillium striatum* Raper & Fennell).

12. **Eupenicillium** Ludwig - Lehrb. nied. Krypt. p. 256 (1892)
    E. crustaceum Ludwig (Fig. 44)
    = *Carpenteles* Langeron - C. r. Séanc. Soc. Biol. *87*: 344 (1922)
    C. glaucum Langeron = *E. crustaceum*
    conidial state: *Penicillium*
    about 30 species
    References: Stolk and Scott, 1967; Scott and Stolk, 1967; Scott, 1968.

The ascomycete genera *Dichlaena* Dur. & Mont., *Penicilliopsis* Solms-Laubach and *Trichocoma* Jungh. (non DC.) = *Trichoskytale* Corda also have *Penicillium*- or *Aspergillus*-conidial states. In pure culture, however, the ascigerous states have not yet been observed.

13. **Sagenoma** Stolk & Orr - Mycologia (in press)
    S. viride Stolk & Orr
    1 species from soil.

14. **Thermoascus** Miehe - Die Selbsterhitzung des Heues p. 70 (1907)
    Th. aurantiacus Miehe
    = *Dactylomyces* Sopp - Skr. Vidensk. Selsk. Christiania, math.-nat. Kl. *11*: 35 (1912)
    D. thermophilus Sopp = *Th. thermophilus* (Sopp) v. Arx
    conidial state: *Paecilomyces*- or *Polypaecilum*-like, all species are thermophilic
    third species: *Th. crustaceus* (Apinis & Chesters) Stolk
    References: Apinis, 1967; Stolk, 1965.

15. **Aphanoascus** Zukal - Ber. dt. bot. Ges. *8*: 295 (1890)
    A. cinnabarinus Zukal (Fig. 49)
    1 keratinophilic species

References: Apinis, 1968; de Vries, 1969; Udagawa and Takada, 1973.

6. **Dichotomomyces** Saito ex Scott - Trans. Br. mycol. Soc. *55*: 314 (1970)
   D. cejpii (Milko) Scott (Fig. 45)
   conidial state: *Polypaecilum*
   1 species.

7. **Leucothecium** v. Arx & Samson - Persoonia 7: 378 (1973)
   L. emdenii v. Arx & Samson
   1 species with a conidial state forming 1-celled, hyaline arthroconidia.

8. **Onygena** Pers. ex Fr. - Syst. Mycol. *3*: 206 (1829)
   O. equina (Willdenow) Pers. ex Fr.
   2 or 3 species, keratinophilic
   References: Samson and van der Aa, 1973.

9. **Xynophila** Malloch & Cain - Can. J. Bot. *49*: 845 (1971)
   X. mephitalis Malloch & Cain
   1 species, probably close to *Arachniotus*.

10. **Arachniotus** Schroet. in Cohn's Krypt. Fl. Schles. *3* (2): 210 (1893)
    A. ruber (van Tieghem) Schroet. (Fig. 46b)
    = *Petalosporus* Ghosh & al. - Mycopath. Mycol. appl. *21*: 36 (1936)
    P. nodulosus Ghosh & al. = *A. citrinus* Massee & Salmon
    = *Pseudoarachniotus* Kuehn - Mycologia *49*: 694 (1957)
    P. roseus Kuehn = *Arachniotus dankaliensis* (Castellani) van Beyma
    = *Waldemaria* Batista & al. - Atas Inst. Micol. (Recife) *1*: 5 (1960)
    W. pernambucensis Batista & al. = *A. dankaliensis*
    6 species, mostly without any conidial states (Fig. 46a)
    References: Apinis, 1964; von Arx, 1971.

11. **Gymnoascus** Baranetzky - Bot. Ztg. *30*: 158 (1872)
    G. reessii Baranetzky (Fig. 46c)
    = *Neogymnomyces* Orr - Can. J. Bot. *48*: 1061 (1970)
    G. demonbreunii Ajello & Cheng
    conidial state: *Chrysosporium*-like
    4 species
    References: Apinis, 1964; Orr, Kuehn and Plunkett, 1963; Samson, 1972.

12. **Pectinotrichum** Varsavsky & Orr - Mycopath. Mycol. appl. *43*: 229 (1971)
    P. llanense Varsavsky & Orr

13. **Narasimhella** Thirum. & Mathur - Sydowia *19*: 184 (1966)
    N. poonensis Thirum. & Mathur
    second, probably indentical species: *N. hyalinospora* (Kuehn & al.) v. Arx
    other species: *Narasimhella echinulata* (Dutta & Ghosh) v. Arx *comb. nov.* = *Pseudoarachniotus echinulatus* Dutta & Ghosh (Mycologia 55: 775, 1963) with oblate, furrowed, echinulate ascospores
    References: von Arx, 1971.

14. **Nannizzia** Stockdale - Sabouraudia *1*: 45 (1961)
    N. gypsea (Nannizzi) Stockdale

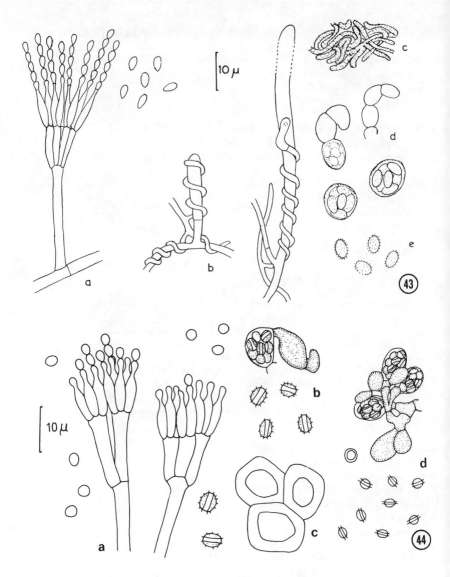

Fig. 43. *Talaromyces flavus*, a, *Penicillium* conidial state; b, ascogonia and antheridia; c, peridial hyphae: d, young and ripe asci; e. ascospores (orig.).

Fig. 44. *Eupenicillium crustaceum*, a, *Penicillium* conidial state; b, asci in chains and ascospores; c, thick-walled peridial cells; d, *Eupenicillium shearii* Stolk & Scott, asci in clusters and ascospores (from Stolk and Scott, 1967).

Fig. 45. *Dichotomomyces cejpii*, a, *Polypaecilum* conidial state; b, ascogonial coil; c, peridial wall; d, asci and ascospores (orig.).

Fig. 46. a, *Arachniotus citrinus* Massee & Salmon, ascogonial coils, asci and ascospores; b, *Arachniotus ruber* (van Tieghem) Schroet., ascogonial coils, asci and ascospores; c, *Gymnoascus reessii* Baranetzki, peridial hyphae with setae, asci and ascospores; d, *Pseudogymnoascus roseus* Raillo, part of an ascoma, peridial hyphae, asci, ascospores and conidia (from Domsch and Gams, 1970, a orig.).

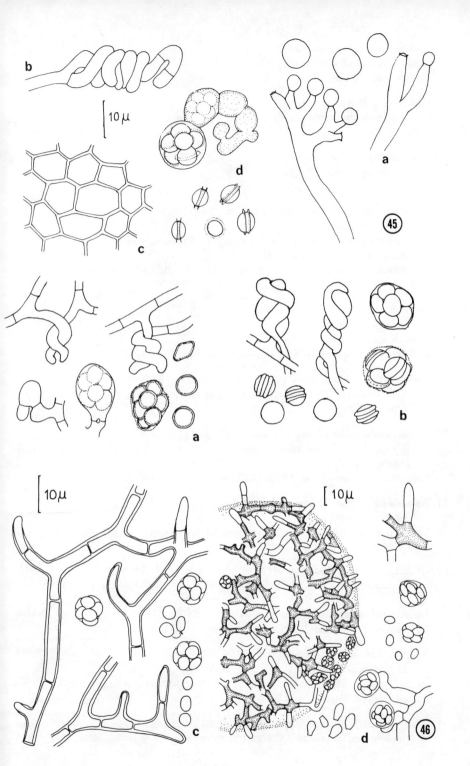

97

conidial state: *Microsporon*
6 or 7 species, all dermatophytic
References: Apinis, 1964.

25. **Shanorella** Benjamin - Aliso *3*: 319 (1956)
    S. spirotricha Benjamin
    1 species.

26. **Spiromastix** Kuehn & Orr - Mycologia *54*: 160 (1962)
    S. warcupii Kuehn & Orr
    1 species.

27. **Arthroderma** Berk. - Outl. Brit. Fungi p. 357 (1860)
    A. curreyi Berk.
    conidial state: *Trichophyton, Keratinomyces, Chrysosporium*
    13 species, mostly dermatophytic
    References: Dawson and Gentles, 1961; Dawson, 1963; Padhye and Carmichael, 1971.

28. **Ctenomyces** Eidam in Cohn - Beitr. Biol. Pfl. *3*: 274 (1880)
    C. serratus Eidam
    1 species
    References: Benjamin, 1956; Orr and Kuehn, 1963.

29. **Ajellomyces** McDonough & Lewis - Mycologia *60*: 77 (1968)
    A. dermatitidis McDonough & Lewis is the ascigerous state of *Blastomyces dermatitidis*.

30. **Emmonsiella** Kwon-Chung - Science *177*: 368 (1972)
    E. capsulata Kwon-Chung is the ascigerous state of *Histoplasma capsulatum* Darling
    References: Kwon-Chung, 1973; Glick and Kwon-Chung, 1973.

31. **Apinisia** LaTouche - Trans. Br. mycol. Soc. *51*: 283 (1968)
    A. graminicola
    The genera *Ajellomyces* and *Emmonsiella* are closely related to *Apinisia* and hardly to distinguish.

32. **Auxarthron** Orr & Kuehn in Orr & al. - Can. J. Bot. *41*: 1439 (1936)
    A. californiense Orr & Kuehn (= *Gymnoascus californiensis* (Orr & Kuehn) Apinis)
    7 partly indistinguishable species are described (Fig. 47)
    References: Domsch and Gams, 1970; Samson, 1972.

33. **Arachnotheca** v. Arx - Persoonia *6*: 376 (1971)
    A. glomerata (Müller & Pacha-Aue) v. Arx
    second species: *A. albicans* (Apinis) v. Arx *comb. nov.* = *Arachniotus albicans* Apinis - Mycol. Pap. *96*: 45 (1964).

34. **Amauroascus** Schroet. - Krypt.-Fl. Schles. *3* (2): 211 (1893)
    Amauroascus verrucosus (Eidam) Schroet.

5 species
References: von Arx, 1971.

5. **Byssoascus** v. Arx - Persoonia *6*: 376 (1971)
    B. striatisporus (Barron & Booth) v. Arx
    conidial state: *Oidiodendron*
    References: Barron and Booth, 1966.

6. **Pseudogymnoascus** Raillo - Zentbl. Bakt. Parasitkde, Abt. 2, *78*: 520 (1929)
    P. roseus Raillo (Fig. 46d)
    conidial state: *Chrysosporium*-like
    second species: *P. bhattii* Samson
    References: Samson, 1972.

7. **Myxotrichum** Kunze - Mykol. Hefte *2*: 109 (1823)
    M. chartarum Kunze (Fig. 48)
    = *Actinospora* Corda - Icon. Fung. *6*: 7 (1854)
    A. chartarum (Nees) Corda = *M. chartarum*
    = *Eidamella* Matr. & Dassonville - Bull. Soc. mycol. Fr. *17*: 129 (1901)
    E. spinosa Matr. & Dassonville = *M. deflexum* Berk.
    = *Toxotrichum* Orr & Kuehn - Mycologia *56*: 473 (1964)
    T. cancellatum (Phillips) Orr & Kuehn = *M. cancellatum* Phillips
    about 10 species
    References: Orr, Kuehn and Plunkett, 1963.

8. **Albertiniella** Kirschst. - Ann. mycol. *34*: 183 (1936)
    A. reticulata Kirschst. = *A. polyporicola* (Jacz.) Malloch & Cain
    1 species
    References: Petrak, 1947; Malloch and Cain, 1972.

9. **Fragosphaeria** Shear - Mycologia *15*: 124 (1923)
    F. purpurea Shear
    second species: *F. reniformis* (Sacc. & Therry) Malloch & Cain.

10. **Cryptendoxyla** Malloch & Cain - Can. J. Bot. *48*: 1816 (1970)
    C. hypophloia Malloch & Cain
    1 species.

11. **Hapsidospora** Malloch & Cain - Can. J. Bot. *48*: 1819 (1970)
    H. irregularis Malloch & Cain
    1 species.

12. **Nigrosabulum** Malloch & Cain - Can. J. Bot. *48*: 1882 (1970)
    N. globosum Malloch & Cain (Fig. 50)
    1 species.

13. **Emericellopsis** van Beyma - Antonie van Leeuwenhoek *6*: 263 (1939)
    E. terricola van Beyma (Fig. 51)
    = *Cypsulotheca* Kamyschko - Not. Syst. crypt. Inst. Bot. Acad. Sci. USSR. *13*: 162 (1960)
    C. aspergilloides Kamyschko = *E. terricola*
    = *Peyronellula* Malan - Mycopath. Mycol. appl. *6*: 164 (1952)
    P. mirabilis Malan = *E. mirabilis* (Malan) Stolk

= *Saturnomyces* Cain - Can. J. Bot. *34*: 135 (1956)
    S. humicola Cain = *E. minima* Stolk
conidial state: *Acremonium*
7 species
References: Backus and Orpurt, 1961; Durrell, 1959; Stolk, 1955; Grosklags and Swift, 1957; Gams, 1970.

44. **Anixiopsis** Hansen - Bot. Ztg. *7*: 131 (1897)
    A. stercoraria Hansen = *A. fulvescens* (Cooke) de Vries (Fig. 52)
    References: Cain, 1957; de Vries, 1969.

45. **Keratinophyton** Randhawa & Sandhu - Sabouraudia *3*: 252 (1964)
    K. terreum Randhawa & Sandhu
    1 species, keratinophilic, with *Chrysosporium*-like conidial state.

46. **Xanthothecium** v. Arx & Samson - Persoonia 7: 377 (1973)
    X. peruvianum (Cain) v. Arx & Samson = *Anixiopsis peruviana* Cain
    1 species.

47. **Arachnomyces** Massee & Salmon - Copr. Fung. *2*: 68 (1902)
    A. nitidus Massee & Salmon
    3 species
    References: Malloch and Cain, 1970.

48. **Pseudeurotium** van Beyma - Zentbl. Bakt. Parasitkde, Abt. 2, *96*: 415 (1937)
    P. zonatum van Beyma (Fig. 53)
    = *Levispora* Routien - Mycologia *49*: 189 (1957)
    L. terricola Routien
    2 or 3 species
    References: Stolk, 1955; Booth, 1961; Mouchacca, 1971.

49. **Xylogone** v. Arx & Nilsson - Svensk. bot. Tidskr. *63*: 345 (1969)
    X. sphaerospora v. Arx & Nilsson (Fig. 54)
    1 species with a conidial state forming septate arthroconidia.

    Mycogala marginata Crooks (1935) may be a similar fungus, but its description partly is based on an accompanying *Fusarium* (which only was developed on a subculture of the type).

50. **Pleuroascus** Massee & Salmon - Ann. Bot., Lond. *15*: 330 (1901)
    P. nicholsonii Massee & Salmon
    1 species
    References: Malloch and Benny, 1973 (Mycologia *65*: 648)

51. **Monascus** van Tieghem - Bull. Soc. bot. Fr. *31*: 266 (1884)
    M. ruber van Tieghem (Fig. 39)
    = *Allescheria* Sacc. & Syd. - Syll. Fung. *14*: 464 (1899)
    A. gayoni (Cost.) Sacc. & Syd. = *Monascus ruber*
    = *Backusia* Thirum. & al. - Mycologia *56*: 813 (1964)
    B. terricola Thirum. & al. = *Monascus ruber*
    = *Microeurotium* Ghatak - Ann. Bot., Lond. *50*: 869 (1936)
    M. albidum Ghatak

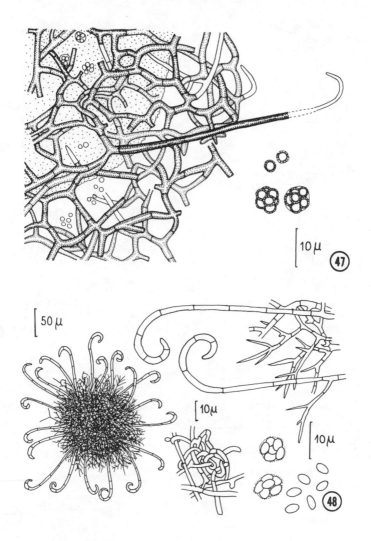

Fig. 47. *Auxarthron umbrinum* (Boud.) Orr & Plunkett, part of an ascoma with peridial hyphae and an appendage, asci and ascospores (fróm Domsch and Gams, 1970).

Fig. 48. *Myxotrichum chartarum*, ascoma, ascocarp initial, appendages, asci and ascospores (from Udagawa, 1962).

= *Xeromyces* Fraser - Proc. Linn. Soc. N.S. Wales 78: 245 (1953)
    X. bisporus Fraser = *M. bisporus* (Fraser) v. Arx

about 15 species. The genus may be related to *Eremascus*

References: Cole and Kendrick, 1958 (esp. for conidial state).

2. **Ascosphaera** Olive & Spiltoir - Mycologia 47: 242 (1955)
    A. apis (Maassen ex Clausen) Olive & Spiltoir

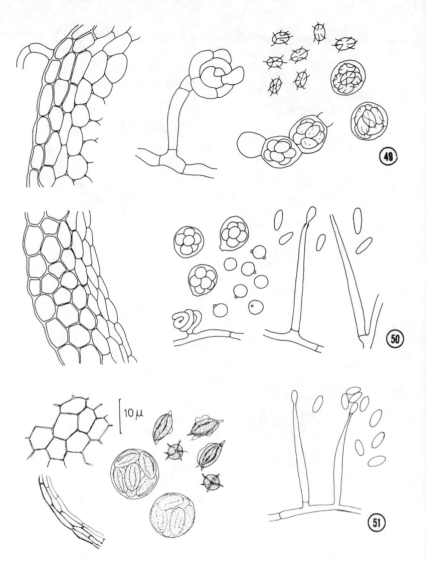

Fig. 49. *Aphanoascus cinnabarinus*, part of the ascocarp wall, initial coil, asci and ascospores (orig.). Fig. 50. *Nigrosabulum globosum*, part of the ascocarp wall, initial coil, asci, ascospores and *Acremonium* conidial state (orig.). Fig. 51. *Emericellopsis terricola*, parts of the ascocarp wall, asci, ascospores and *Acremonium* conidial state (orig.).

3 species on bees
References: Skou, 1972.

53. **Bettsia** Skou - Friesia *19* (1): 5 (1972)
      B. alvei (Betts) Skou = *Ascosphaera alvei* (Betts) Olive & Spiltoir
= *Pericystis* Betts - Ann. Bot. *26*: 795 (1912) non Agardt (1848)
1 osmophilic species (compare also *Monascus* and *Eremascus*).

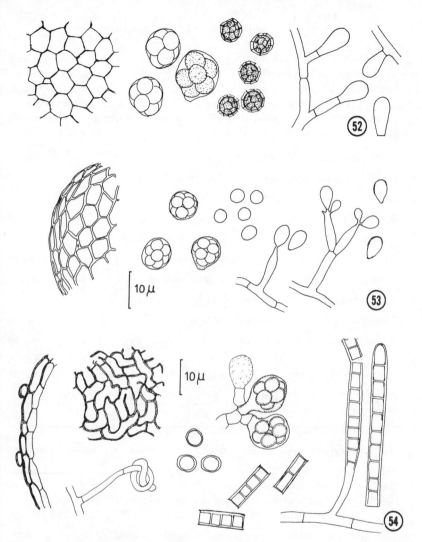

Fig. 52. *Anixiopsis fulvescens*, part of the ascocarp wall, asci, ascospores and conidia (aleuriospores) (orig.). Fig. 53. *Pseudeurotium zonatum*, part of the ascocarp wall, asci, ascospores and conidial state (orig.). Fig. 54. *Xylogone sphaerospora*, parts of the ascocarp wall, ascocarp initial, asci, ascospores and conidial state with endogenous arthrospores (orig.).

. **Amorphotheca** Parbery - Ast. J. Bot. *17*: 342 (1969)
    A. resinae Parbery
conidial state: *Cladosporium resinae* (Lindau) de Vries = *Hormoconis resinae* (Lindau) v. Arx & de Vries.

. **Ephemeroascus** van Emden - Trans Br. mycol. Soc. *61*: 599 (1973)
    E. verticillatus van Emden
1 species with a *Verticillium* conidial state.

SPHAERIALES

Only representatives of relatively few genera of this voluminous order of Ascomycetes develop ascomata with ripe asci in pure culture. With many species attempts at culturing have been either unsuccessful or only sterile mycelium or the conidial state was obtained. Nearly all known genera of Sphaeriales are enumerated in the keys given by Müller and von Arx (1973). Especially genera known to include species studied in pure culture can be distinguished by the following key:

1. Apical structures of the asci not amyloid 2

1. Apical structures of the asci at least partly amyloid 7

2. Asci without refractive apical rings or caps 3

2. Asci with refractive apical rings or caps 6

3. Ascospores without germpores or germslits (furrows) 4

3. Ascospores with (often inconspicuous) germpores or germslits or apiculate 10

4. Ascomata dark, membranaceous or carbonaceous 5

4. Ascomata fleshy, light, yellow, red, green or blue 47

5. Asci small, spherical, evanescent, ascospores 1-celled, small, hyaline 8

5. Asci persistent, mostly elongated 60

6. Asci with an apical ring, usually visible as 2 refractive bodies, ascospores mostly not ejaculated, ascomata often with a long neck 68

6. Asci with an apical cap or body, perforated by a narrow canal, ascospores often fusiform, apiculate or filiform 78

7. Ascospores not allantoid or reniform, without a germslit, often septate . 83

7. Ascospores allantoid or reniform, 1-celled, pigmented, often with a germslit 91

8. Asci evanescent, ascospores with a slimy sheath, conidial state is *Chalara* (*Chalaropsis, Thielaviopsis*) with endogenous, cylindrical phialoconidia . . . . *Ceratocystis* (1)

8. Asci indistinct, without membrane, ascospores embedded in a tough slimy mass, conidial state is *Sporothrix, Verticicladiella* or *Graphium* with small blastoconidia formed in sympodulae 9

9. Ascomata ostiolate . . . . *Ophiostoma* (2)

9. Ascomata non-ostiolate . . . . *Europhium* (3)

10. Ascospores with a germslit or furrow, 1-celled 11
(For genera with septate ascospores compare Pseudosphaeriales)

10. Ascospores with germpores 15

11. Ascomata develop on an erumpent stroma, asci 4-spored . . . . . . *Wawelia* (6)

11. Ascomata discrete or crustose 12

12. Ascomata non-ostiolate 13

(and many other genera of Xylariaceae, Diatrypaceae, Amphisphaeriaceae, Diaporthaceae, etc., not yet known in pure culture.)

List of genera

1. **Ceratocystis** Ellis & Halst. - Bull. N.J. agric. Exp. Stn 76: 14 (1890)
    C. fimbriata Ellis & Halst. (Fig. 55)
= *Endoconidiophora* Münch - Nat. Z. Forst- u. Landw. 6: 34 (1908)
    E. coerulescens Münch = *C. coerulescens* (Münch) Bakshi
= *Rostrella* Zimmerm. - Meded. s'Lands Plantentuin 37: 24 (1900) non Fabre (1878)
    R. coffeae Zimmerm. = *C. fimbriata*
conidial state: *Thielaviopsis, Chalara, Chalaropsis*
10 species
References: Hunt, 1956; Davidson, 1958, 1964; Hinds and Davidson, 1967; Griffin, 1968.

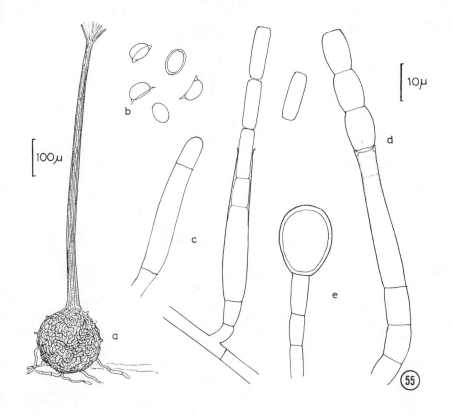

Fig. 55. *Ceratocystis fimbriata*, a, perithecium; b, ascospores; c, d, *Chalara* conidial state (endoconidia); e, aleuriospore (chlamydospore) (orig.).

2. **Ophiostoma** Syd. - Ann. mycol. *17*: 43 (1919)
   O. pilifera (Fr.) Syd.
   = *Linostoma* Höhnel - Ann. mycol. *16*: 91 (1918) non Wallich (1831)
      L. pilifera (Fr.) Höhnel = *O. pilifera*
   = *Grosmannia* Goidanich - Boll. R. Staz. Pat. veg. Roma *16*: 26 (1936)
      G. serpens Goidanich = *O. serpens* (Goidanich) v. Arx
   conidial states: *Sporothrix, Pesotum, Graphium, Verticicladiella*; most of the
   species comprise a *Sporothrix-* and a *Verticicladiella-* or *Pesotum/Graphium--*
   conidial form.
   about 50 species
   References: see *Ceratocystis*.

3. **Europhium** Parker - Can. J. Bot. *35*: 175 (1957)
   E. trinacriforme Parker
   conidial state: *Verticicladiella*

111

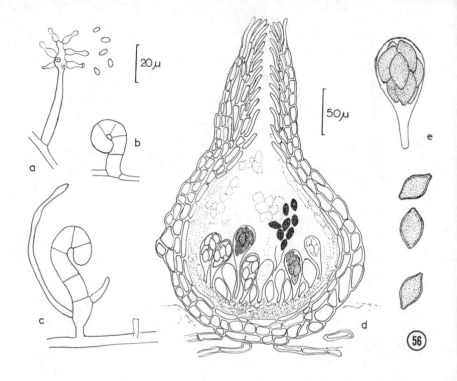

Fig. 56. *Melanospora zamiae*, a, microconidia, borne on phialides; b, c, ascogonia; d, perithecium; e, ascus and ascospores (from Doguet, 1955).

4 species. *Europhium* is the non-ostiolate counterpart of *Ophiostoma*. In the related genus *Cephaloascus* the asci are formed on *Verticicladiella*-like ascophores. References: Robinson-Jeffrey and Davidson, 1968.

4. **Sphaeronaemella** Karst. - Hedwigia *23*: 17 (1884)
    S. helvellae Karst.
second species: *S. fimicola* Marchal
References: Seeler, 1943; Cain and Weresub, 1957.

5. **Melanospora** Corda - Icon. Fung. *1*: 24 (1837)
    M. zamiae Corda (Fig. 56)
= *Ampullaria* A. L. Smith - J. Bot. *61*: 258 (1903)
    M. aurea A. L. Smith = *M. zamiae*
= *Ceratostoma* Fr. - Summa veg. Scand. p. 396 (1849)
    C. chioneum Fr. = *M. chionea* (Fr.) Corda
= *Gibsonia* Massee - Ann. Bot. *23*: 336 (1909)
    G. phaeospora Massee = *M. zamiae*

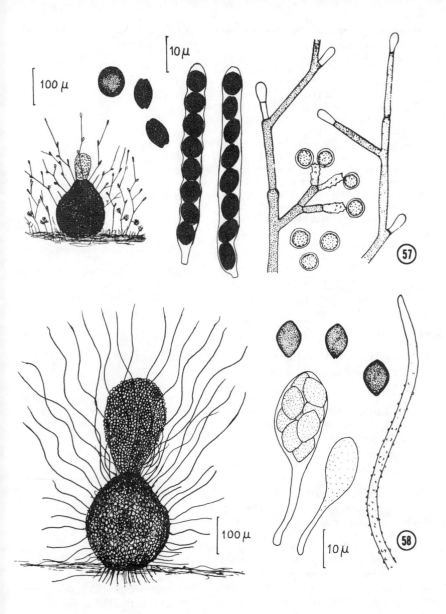

Fig. 57. *Ascotricha chartarum*, perithecium, asci, ascospores and conidial state with conidia borne on denticles (orig.).
Fig. 58. *Chaetomium globosum*, perithecium, asci and ascospores (orig.).

= *Guttularia* Obermeyer - Mycol. Zentbl. *3*: 9 (1913)
    G. geosporae Obermeyer = *M. geosporae* (Obermeyer) Höhnel
= *Lithomyces* Viala & Marsais - Ann. Inst. Nat. Rech. agron. *23*: 188 (1930)
    L. nidulans Viala & Marsais = *M. zobelii*

113

= *Melanosporopsis* Naumov - Mat. Mik. Fitopat. *6*: 6 (1927)
  M. subulata Naumov
= *Microthecium* Corda - Icon. Fung. *5*: 30 (1842)
  M. zobelii Corda = *M. zobelii* (Corda) Fuckel
= *Nigrosphaeria* Gardner - Publ. Univ. Calif. Bot. *2*: 191 (1905)
  N. setchellii (Harkn.) Gardner = *M. setchellii* (Harkn.) Sacc. & Syd.
= *Rhynchomelas* Clem. - Gen. Fungi p. 173 (1909)
  R. arenariae (Mont.) Clem. = *M. arenariae* Mont.
= *Scopinella* Lév. - Dict. Univ. W. L. *8*: 493 (1849)
  S. barbata (Pers.) Lév. = *M. barbata* (Pers. ex Fr.) Dur. & Mont.
= *Sphaeroderma* Fuckel - Symb. mycol., Nachtr. *3*: 23 (1872)
  S. theleboloides Fuckel = *M. theleboloides* (Fuckel) Wint.
= *Sphaerodes* Clem. - Gen. Fungi p. 173 (1909)
  S. episphaerium (Phill. & Plowr.) Clem. = *M. episphaeria* Phill. & Plowr.
= *Vittadinula* Sacc. - Syll. Fung. *24*: 650 (1926)
  V. episphaeria (Phill. & Plowr.) Sacc. = *M. episphaeria*
about 50 species
References: Doguet, 1955; Udagawa and Cain, 1969 (sub *Microthecium* for non-ostiolate species).

6. **Wawelia** Namyslowski - Bull. Acad. Sci. Cracovie: 567 (1908)
   W. regia Namyslowski
   1 species
   References: Müller, 1959.

7. **Phaeostoma** v. Arx & Müller - Beitr. Krypt. Fl. Schweiz *11*(1): 148 (1954)
   P. vitis (Fuckel) v. Arx & Müller
   1 species.

8. **Ascotricha** Berk. - Ann. Mag. nat. Hist. *1*: 257 (1838)
   A. chartarum Berk. (Fig. 57)
   6 or 7 species
   conidial state: *Dicyma* Boul.
   References: Ames, 1961 (1963); Hawksworth, 1971.

9. **Chaetomium** Kunze ex Fr. - Syst. mycol. *3*: 225 (1829)
   Ch. globosum Kunze ex Fr. (Fig. 58)
   = *Bolacotricha* Berk. & Br. - Ann. Mag. nat. Hist., Ser. *2*,12: 97 (1851)
     B. grisea Berk. & Br. = *Ch. murorum* Corda
   = *Bommerella* March. - Bull. Soc. Roy. Belg.: 1 (1885)
     B. trigonospora March. = *Ch. trigonosporum* (March.) Chivers
   about 100 species, often cellulolytic
   References: Ames, 1961 (1963); Skolko and Groves, 1948, 1953; Seth, 1967, 1970; Lodha, 1964; Udagawa, 1960, 1965; Mazzucchetti, 1965; Gams, 1966; Udagawa and Cain, 1969.

10. **Chaetomidium** (Fuckel) Zopf - Nova Acta Leop.-Carol. *42*: 199 (1881)
    Ch. fimeti (Fuckel) Zopf
    6 or 7 species
    References: Cain, 1961; Seth, 1967, 1968.

11. **Achaetomium** Rai, Tewari & Mukerji - Can. J. Bot. *42*: 693 (1964)
    A. globosum Rai & Tewari

second species: *A. strumarium* Rai, Tewari & Mukerji
(*A. luteum* Rai & Tewari is an uncertain species and seems to belong to another genus, probably to *Achaetomiella*).

2. **Microascus** Zukal - Verh. zool. bot. Ges. Wien *35*: 333 (1885)
    M. longirostris Zukal (Fig. 59)
  = *Nephrospora* Loubière - C-r. Paris *177*: 211 (1923)
    N. manginii Loubière = *M. manginii* (Loubière) Curzi
  = *Peristomium* Lechmère - Bull. Soc. mycol. Fr. *29*: 307 (1913)
    P. desmosporum Lechmère = *M. desmosporum* (Lechmère) Curzi
  conidial state: *Scopulariopsis, Wardomyces*
  10 species
  References: Barron, Cain and Gilman, 1961; Udagawa, 1962, 1963; Doguet, 1957; Morton and Smith, 1963; Malloch, 1970.

3. **Kernia** Nieuwland - Am. Midl. Nat. *4*: 379 (1916)
    K. nitida (Sacc.) Nieuwland
  ≡ *Magnusia* Sacc. (1875) non *Magnusia* Klotzsch (1854)
  conidial state: *Scopulariopsis, Graphium*
  4 or 5 species
  References: Benjamin, 1955, 1956; Seth, 1968; Malloch and Cain, 1971.

4. **Pithoascus** v. Arx - Proc. K. ned. Akad. Wet. C. *76*: 295 (1973)
    P. nidicola (Massee & Salmon) v. Arx
  6 species
  References: von Arx, 1973.

5. **Lophotrichus** Benjamin - Mycologia *41*: 347 (1949)
    L. ampullus Benjamin
  4 species
  References: Ames, 1961 (1963); Malloch and Cain, 1971; Seth, 1971.

6. **Petriella** Curzi - Boll. Staz. Pat. Veg. Roma *10*: 380 (1930)
    P. asymmetrica Curzi = *P. sordida* (Zukal) Barron & Gilman
  conidial state: '*Graphium*' or '*Sporocybe*', mostly with synnemata and with 1-celled, hyaline conidia developed sympodially by budding with a wide base.
  5 species (Fig. 60)
  References: Barron, Cain and Gilman, 1961; Udagawa, 1963.

7. **Petriellidium** Malloch - Mycologia *62*: 738 (1970)
    P. boydii (Shear) Malloch
  conidial states: *Scedosporium* Sacc., *Graphium*, and others
  4 or 5 species
  References: von Arx, 1973.

8. **Thielavia** Zopf - Verh. bot. Ver. Brandenburg *18*: 101 (1876)
    T. basicola Zopf
  = *Boothiella* Lodhi & Mirza - Mycologia *54*: 217 (1962)
    B. tetrasperma Lodhi & Mirza = *Thielavia tetrasperma* (Lodhi & Mirza) v. Arx *comb. nov.*
  = *Thielaviella* v. Arx & Tariq - Trans. Br. mycol. Soc. *51*: 611 (1968)
    T. humicola v. Arx & Tariq = *Thielavia tetrasperma*

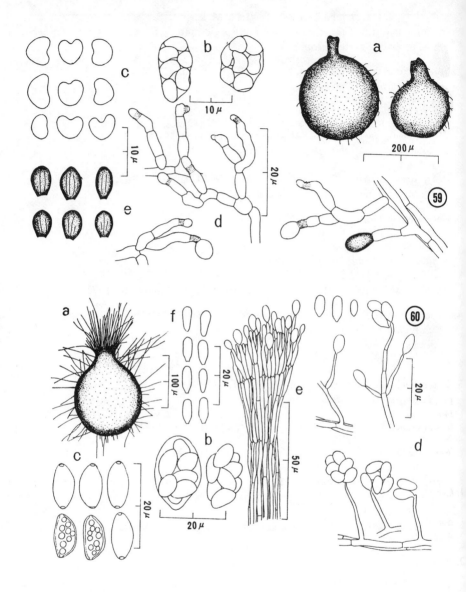

Fig. 59. *Microascus doguetii* Moreau, a, perithecia; b, asci; c, ascospores; d, *Scopulariopsis* conidial state; e, conidia (from Udagawa, 1963).

Fig. 60. *Petriella setifera* (Schmidt) Curzi, a, perithecium; b, asci; c, ascospores; d, e, *Graphium* conidial state; f, conidia (from Udagawa, 1963).

about 12 species

References: Booth, 1961; Booth and Shipton, 1966; Doguet, 1956; Lucas, 1949.

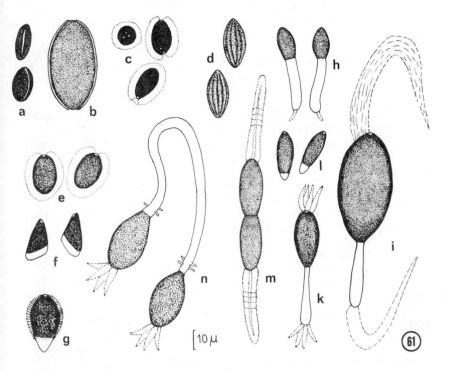

Fig. 61. Ascospores of: a, *Coniochaeta tetraspora* Cain; b, *Bombardioidea bombardioides*; c, *Fimetariella rabenhorstii*; d, *Neurospora sitophila*; e, *Sordaria fimicola*; f, *Triangularia bambusae*; g, *Apiosordaria verruculosa*; h, *Cercophora lignicola*; i, *Podospora fimiseda* (Ces. & de Not.) Niessl; k, *Podospora vestita* (Zopf) Wint.; 1, *Zopfiella* spec. (*Entosordaria vestita* (Zopf) Wint.); m, *Zygospermella insignis*; n, *Zygopleurage zygospora*.

). **Corynascus** v. Arx - Proc. K. ned. Akad. Wet. C. 76: 295 (1973)
   C. sepedonium (Emmons) v. Arx = *Thielavia sepedonium* Emmons
   conidial state: *Sepedonium, Chrysosporium*
   4 species.

). **Achaetomiella** v. Arx - Gen. Fung. p. 247 (1970)
   A. virescens v. Arx
   2 species. The genus is related to *Corynascus*.

l. **Coniochaeta** (Sacc.) Massee - Grevillea 16: 37 (1887)
   C. ligniaria (Grev.) Massee
   = *Coniomela* (Sacc.) Kirschst. - Trans. Br. mycol. Soc. 18: 306 (1933)
      C. pulveracea (Ehrh.) Kirschst. = *Coniochaeta pulveracea* (Ehrh.) Munk
   = *Cucurbitariella* Petrak - Ann. mycol. 14: 440 (1916)
      C. moravica Petrak
   = *Sphaerodermella* Höhnel - Sber. K. Akad. Wiss. Wien, math.-nat. Kl. 1,116: 105 (1907)
      S. niesslii (Auersw.) Höhnel = *Coniochaeta niesslii* (Auersw.) v. Arx & E. Müller

conidial state: *Phialophora*
about 15 species (Fig. 61a)
References: von Arx and Müller, 1954; Cain, 1934; Munk, 1957; Udagawa and Takada, 1967.

22. **Coniochaetidium** Malloch & Cain - Can. J. Bot. *49*: 878 (1971)
    C. ostreum Malloch & Cain
    second species: *C. savoryi* (Booth) Malloch & Cain.

23. **Bombardia** Fr. - Summa veg. Scand. p. 389 (1849)
    B. fasciculata (Batsch) Fr.
    1 species (the species enumerated by Cain, 1934, have to be arranged within *Cercophora*).
    References: von Arx and Müller, 1954.

24. **Cercophora** Fuck. - Symb. mycol. p. 244 (1869)
    C. mirabilis Fuck.
    = *Lasiosordaria* Chenant. - Bull. Soc. mycol. Fr. *35*: 77 (1919)
      L. lignicola (Fuckel) Chenant. = *C. caudata* (Curr.) Lundq.
    about 15 species (Fig. 61h)
    References: Lundqvist, 1972.

25. **Apiosordaria** v. Arx & W. Gams - Nova Hedwigia *13*: 201 (1966)
    A. verruculosa (Jensen) v. Arx & W. Gams (Fig. 61g)
    conidial state: *Cladorrhinum*
    1 species.

26. **Triangularia** Boedijn - Ann. mycol. *32*: 302 (1934)
    T. bambusae (van Beyma) Boedijn (Fig. 61f)
    = *Trigonia* van Beyma (1933) non *Trigonia* Aubl.
    4 or 5 species
    References: Cain and Farrow, 1956; von Arx and Hennebert, 1969.

27. **Zopfiella** Wint. in Rabenh. - Krypt. Fl. *1*(2): 56 (1887)
    Z. tabulata (Zopf) Wint.
    = *Strattonia* Cif. - Sydowia *8*: 245 (1954)
      S. tetraspora (Stratton) Cif.
    = *Tripterospora* Cain - Can. J. Bot. *34*: 700 (1956)
      T. longicaudata Cain = *Z. longicaudata* (Cain) v. Arx
    about 5 species, but the genus needs revision (Fig. 61b).

28. **Echinopodospora** Robison - Trans. Br. mycol. Soc. *54*: 318 (1970)
    E. jamaicensis Robison
    2 or 3 species, closely related to *Zopfiella*.

29. **Lacunospora** Cailleux - Cahiers de la Maboké *6*: 91 (1968)
    L. stercoraria Cailleux.

30. **Anopodium** Lundq. - Bot. Not. *117*: 356 (1964)
    L. ampullaceum Lundq.

31. Sordaria Ces. & de Not. - Comm. Soc. critt. Ital. *1*: 225 (1863)
   S. fimicola (Rob.) Ces. & de Not. (Fig. 61e)
= *Fimetaria* Griff.& Seaver - N. Amer. Fl. *3*: 65 (1910)
   F. fimicola (Rob.) Griff. & Seaver = *S. fimicola*
= *Hansenia* Zopf - Z. Naturw. *56*: 27 (1883)
   H. lanuginosa Zopf
= *Ixodopsis* Karst. - Bidr. Känn. Finl. Nat. Folk *23*: 6 (1873)
   5 or 6 species, mostly on dung or on seed or from soil
   References: Moreau, 1954; Cain, 1934; Cain and Groves, 1948; Cailleux, 1972.

32. Arnium Nits.ap. Fuckel - Symb. mycol., Nachtr. *1*: 38 (1872)
   A. lanuginosum (Preuss) Nits. = *A. olerum* (Fr.) Lundq.
= *Pleurosordaria* Fernier - Rev. Mycol., suppl. colon. p. 17 (1954)
   P. brassicae (Kl.) Fernier = *A. olerum*
   about 15 species
   References: Lundqvist, 1972; Cain and Mirza, 1972.

33. Podospora Ces. - Hedwigia *1*: 103 (1852)
   P. fimicola Cesati = *P. fimiseda* (Ces. & de Not.) Niessl
= *Malinvernia* Rabenh. - Hedwigia *1*: 116 (1852)
   M. anserina Rabenh. = *P. anserina* (Rabenh.) Rehm
= *Pleurage* (Fr.) Kuntze - Rev. Gen. Plant. *3*: 505 (1889)
   P. curvula (de Bary) Kuntze = *P. curvula* (de Bary) Niessl
= *Philocopra* Speg. - An. Soc. scient. Argent. p. 192 (1880)
   about 70 species, mostly isolated from dung (Fig. 61i,k)
   References: Cain, 1962; Moreau, 1954 (sub *Pleurage*); Cain, 1934 (sub *Sordaria*);
   Boedijn, 1962; Mirza and Cain, 1969; Lundqvist, 1972 (as *Podospora* and *Schizo-
   thecium*).

34. Bombardioidea C. Moreau - Sordaria et Pleurage p. 132 (1953)
   B. bombardioides (Auersw.) C. Moreau (Fig. 61b)
   1 species on dung (not known in pure culture).

35. Amphisphaerella (Sacc.) Kirschst. - Trans. Br. mycol. Soc. *18*: 306 (1933)
   A. amphisphaerioides (Sacc. & Speg.) Kirschst.
   2 species
   References: Munk, 1953; von Arx and Müller, 1954.

36. Helminthosphaeria Fuckel - Symb. mycol. p. 166 (1869)
   H. clavariae (Tul.) Fuckel
   1 species, parasitic on *Clavaria*, with a *Diplococcium*-like conidial state.

37. Neurospora Shear & Dodge - J. agric. Res. *34*: 1025 (1927)
   N. sitophila Shear & Dodge (Fig. 61a)
   conidial state: *Monilia*
   9 species
   References: Nelson and Backus, 1968; Mahoney, Huang and Backus, 1969; Frederick & al., 1969; Moreau-Froment, 1956.

38. Gelasinospora Dowding - Can. J. Res., Sect. C, *9*: 294 (1933)
   G. tetrasperma Dowding (Fig. 62)

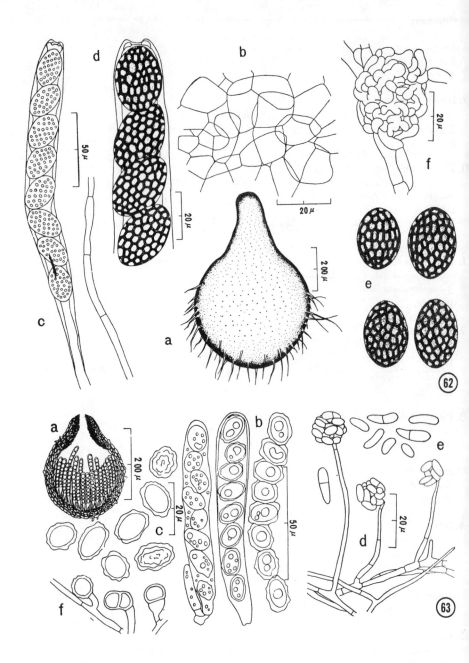

Fig. 62. *Gelasinospora reticulispora* (Greis) C. & M. Moreau, a, perithecium; b, perithecial wall; c, ascus; d, apical portion of an ascus; e, ascospores; f, perithecial initial (from Udagawa, 1966). Fig. 63. *Neocosmospora vasinfecta*, a, perithecium; b, asci; c, ascospores; d, e, conidia borne on phialides; f, chlamydospores (from Udagawa, 1963).

= *Anixiella* Saito & Minoura - J. Ferment. Technol. *26*: 4 (1948)
  A. reticulata (Booth & Ebben) Cain = *C. reticulata* (Booth & Ebben) Cailleux
5 or 6 species (Fig. 62)
References: Cain, 1950, 1961; Udagawa and Takada, 1967; Cailleux, 1972; von Arx, 1973.

9. **Diplogelasinospora** Cain - Can. J. Bot. *39*: 1169 (1961)
  D. princeps Cain
2 species from seed
References: Udagawa and Horie, 1972.

0. **Zygospermella** Cain - Mycologia *27*: 227 (1935)
  Z. setosa Cain = *Z. insignis* (Mouton) Cain (Fig. 61m)
≡*Zygospermum* Cain - Univ. Toronto Stud., Biol. Ser. *38*: 73 (1934) non Thwaites ex Baillon (1858)
second species: *Z. striata* Lundqvist
References: Lundqvist, 1969.

1. **Zygopleurage** Boedijn - Persoonia *2*: 316 (1962)
  Z. zygospora (Speg.) Boedijn (Fig. 61n)
3 species on dung
References: Lundqvist, 1969.

2. **Neocosmospora** E.F. Smith - Bull. U.S. Dept. Agr. Veg. Phys. *17*: 45 (1899)
  N. vasinfecta E. F. Smith (Fig. 63)
further species: *N. africana* v. Arx, *N. ornamentata* Barbosa
conidial state: *Acremonium* (*Cephalosporium*)-like
References: von Arx, 1955; Douget, 1956; Udagawa, 1963.

3. **Pseudonectria** Seaver - Mycologia *1*: 48 (1909)
  P. rousseliana (Mont.) Seaver
≡*Nectriella* Sacc. - Michelia *1*: 51 (1877) non Nitschke (1869)
≡*Notarisiella* (Sacc.) Clem. & Shear - Gen. Fungi p. 280 (1931)
conidial states: *Volutella* (*Volutella buxi* (Corda) Berk.), *Sesquicillium*
5 or 6 species, imperfectly known in culture
References: Bezerra, 1963; Juel, 1925 (as *Nectriella*).

4. **Nectria** Fr. - Summa Veg. Scand. p. 287 (1847)
  N. cinnabarina (Tode ex Fr.) Fr.
synonyms vide Müller and von Arx, 1962
conidial states: *Tubercularia, Cylindrocarpon, Fusarium, Zythia, Stilbella*
about 40 species (Fig. 64)
References: Booth, 1959, 1960; Hanlin 1961, 1963; Dingley, 1951, 1957; Wollenweber and Reinking, 1935 (*Hypomyces* sensu Wollenweber = *Nectria*).

5. **Gibberella** Sacc. - Michelia *1*: 43, 317 (1877)
  G. pulicaris (Fr.) Sacc.
conidial state: *Fusarium*
about 10 species, including *G. fujikuroi* (Saw.) Wollenw. (in use for the production of gibberellin or gibberellic acid) and *G. zeae* (Schw.) Petch.

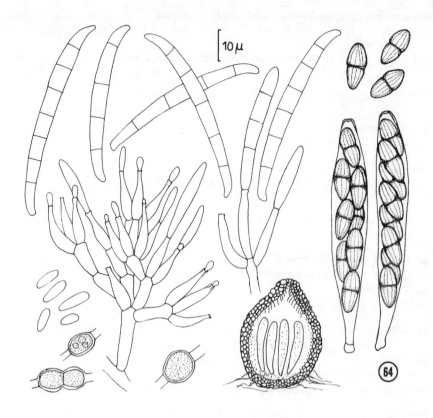

Fig. 64. *Nectria haematococca* Berk. & Br., perithecium, asci, ascospores and conidiophores, phialides and conidia of the *Fusarium* conidial state (*Fusarium javanicum* Koorders) (orig.).

46. **Calonectria** de Not. - Comm. Soc. critt. ital. *2*: 477 (1867)
    C. daldiniana de Not.
conidial states: *Fusarium, Cylindrocladium, Acremonium, Verticillium* and related genera
about 40 species, some parasitic on other fungi
References: Wollenweber and Reinking, 1935; Hansford, 1946.

47. **Scoleconectria** Seaver - Mycologia *1*: 197 (1901)
    S. cucurbitula (Tode ex Fr.) Booth
conidial state: *Zythiostroma* (*Zythia*)
References: Booth, 1959.

48. **Thyronectria** Sacc. - Grevillea *4*: 21 (1875)
    T. patavina Sacc.
= *Pleonectria* Sacc. - Fungi venet. *5*: 178 (1876)
    P. lamyi (Desm.) Sacc. = *T. lamyi* (Desm.) Seeler

conidial state: *Tubercularia*
References: Booth, 1959.

9. **Nectriopsis** Maire - Ann. mycol. *9*: 323 (1911)
    N. violacea (Schmidt ex Fr.) Maire
    second species: *N. candicans* (Plowr.) Maire, both parasitic on Myxomycetes
    References: Samuels, 1973 (Mycologia *65*: 401, as *Nectria*).

0. **Heleococcum** Jørgensen - Bot. Tidskr. *37*: 417 (1922)
    H. auranthiacum Jørgensen
    second species: *H. japonense* Tubaki, 1967.

1. **Roumegueriella** Speg. - Rev. Mycol. *2*: 18 (1880)
    R. murocospora Speg. = *R. rufula* (Berk. & Br.) Malloch & Cain
    = *Lilliputia* Boud. & Pat. - Bull. Soc. mycol. Fr. *16*: 144 (1900)
       L. gaillardii Boud. & Pat. = *R. rufula*
    References: Hughes, 1951; Malloch and Cain, 1972.

2. **Podostroma** Karst. - Hedwigia *31*: 294 (1892)
    P. alutaceum (Pers.) Atk.
    conidial state: *Trichoderma*
    References: Boedijn, 1938.

3. **Hypocrea** Fr. - Syst. Orb. Veget. p. 104 (1825)
    H. rufa (Pers.) Fr.
    for synonyms see Müller and von Arx, 1962
    conidial state: *Trichoderma, Gliocladium, Verticillium*
    References: Rifai and Webster, 1966; Boedijn, 1964; Doi, 1972.

4. **Mycorhynchus** Sacc. - Syll. Fung. *18*: 418 (1906)
    M. betae (Holl.) Sacc.
    = *Rhynchomyces* Sacc. & March. (1892) non Willk. (1866).

5. **Glomerella** Spauld. & v. Schrenck - Bull. U.S. Dept. Agric. *44*: 29 (1903)
    G. cingulata (Stonem.) Spauld. & v. Schrenck (Fig. 65)
    = *Caulochora* Petrak - Ann. mycol. *38*: 341 (1940)
       C. baumgartneri Petrak = *G. cingulata*
    ≡ *Gnomoniopsis* Stonem. - Bot. Gaz. *26*: 99 (1898) non Berlese
    = *Haplothecium* Theiss. & Syd. - Ann. mycol. *13*: 614 (1915)
       H. amenti (Rostr.) Theiss. & Syd. = *G. amenti* (Rostr.) v. Arx & E. Müller
    = *Hypostegium* Theiss. - Verh. zool. bot. Ges. Wien *66*: 384 (1916)
       H. phormii (Speg.) Theiss. = *G. cingulata*
    = *Neozimmermannia* Koorders - Verh. K. ned. Akad. Wet. *13*(4): 68 (1907)
       N. elasticae Koorders = *G. cingulata*
    conidial state: *Colletotrichum* (often described as *Gloeosporium*)
    6 or 7 species, among which *G. tucumanensis* (Speg.) v. Arx & E. Müller, parasitic
    on sugarcane, maize and other cereals and grasses.
    References: von Arx, 1957, 1970; von Arx and Müller, 1954.

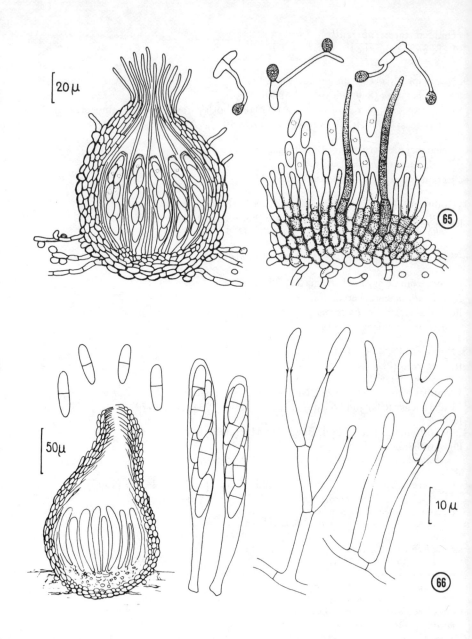

Fig. 65. *Glomerella cingulata*, perithecium, acervulus of the *Colletotrichum* conidial state and germinating conidia (orig.).
Fig. 66. *Plectosphaerella cucumerina*, perithecium, asci, ascospores, conidiophores with phialides and conidia of the *Fusarium* conidial state (orig.).

124

. **Trichosphaerella** Bomm. & al. - Syll. Fung. *9*: 604 (1891)
    T. decipiens Bomm. & al.
= *Bresadolella* Höhnel - Ann. mycol. *1*: 522 (1903)
    B. aurea Höhnel = *T. decipiens*
= *Larseniella* Munk - Bot. Tidskr. *46*:58 (1942)
    L. globulispora Munk = *T. ceratophora* (Höhnel) E. Müller
= *Melanopsammella* Höhnel - Ann. mycol. *17*: 121 (1919)
    M. inaequalis (Grove) Höhnel = *T. inaequalis* (Grove) E. Müller
= *Neorehmia* Höhnel - Sber.K.Akad.Wiss.Wien, math.-nat. Kl. *111*(1): 988 (1902)
    N. ceratophora Höhnel = *T. ceratophora*
= *Oplothecium* Syd. - Ann. mycol. *21*: 97 (1923)
    O. arecae Syd. = *T. arecae* (Syd.) E. Müller
= *Oplotheciopsis* Batista & Cif. - Saccardoa *2*: 243 (1963)
    O. palmae (Stev.) Batista & Cif. = *T. arecae*
4 species
References: Müller and von Arx, 1962; Müller and Dennis, 1965.

. **Niesslia** Auersw. - Mycol. Europaea 5/6: 30 (1869)
    N. exilis (Alb. & Schw.) Wint.
conidial state: *Monocillium*
5 or 6 species
References: Müller and von Arx, 1962; Gams, 1970.

. **Trichosphaeria** Fuckel - Symb. mycol. p. 144 (1869)
    T. pilosa (Pers.) Fuckel
References: von Arx and Müller, 1954.

. **Eriosphaeria** Sacc. - Atti Soc. Veneto-Trent. *4*: 10 (1875)
    E. vermicularia (Nees ex Fr.) Sacc.
References: Müller and von Arx (1962).

. **Plectosphaerella** Kleb. - Phytopath. Z. *1*: 43 (1930)
    P. cucumeris Kleb. = *P. cucumerina* (Lindfors) W. Gams (Fig. 66)
1 species, conidial state *Fusarium tabacinum* (van Beyma) W. Gams
References: Gams and Gerlagh, 1968; Domsch and Gams, 1972.

. **Chaetosphaeria** Tul. - Sel. Fung. Carpol. *2*: 252 (1863)
    Ch. innumera Tul.
synonyms vide Müller and von Arx, 1962
about 10 species
conidial state: *Chloridium, Codinaea, Menispora*
References: Booth, 1957, 1958; Müller and von Arx, 1962.

. **Claviceps** Tul. - Ann. Sci. nat., Bot., Sér. 3, *20*: 43 (1853)
    C. purpurea (Fr.) Tul.
conidial state: *Sphacelia* (*Sphacelia segetum* Lév.).

. **Epichloë** Fr. - Summa Veg. Scand. p. 381 (1849)
    E. typhina (Pers. ex Fr.) Tul.
conidial state: *Sphacelia* (*Sphacelia typhina* Sacc.).

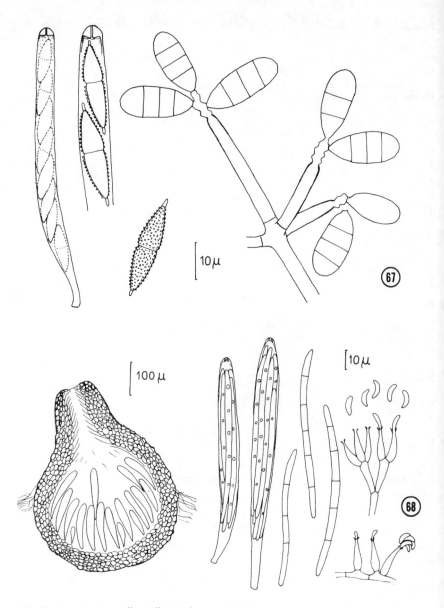

Fig. 67. *Hypomyces rosellus* (Alb. & Schw. ex Fr.) Tul., ascus, apical part of an ascus, ascospore and conidial state (*Cladobotryum dendroides*) (orig.).

Fig. 68. *Gaeumannomyces graminis*, perithecium, asci, ascospores and *Phialophora* conidial state (orig.).

64. **Cordyceps** (Fr.) Link - Handb. Erk. Gewächse *3*: 347 (1833)
    C. militaris (L. ex St. Amans) Link
conidial state: ? *Isaria, Verticillium*.

. Peckiella (Sacc.) Sacc. - Syll. Fung. *9*: 944 (1891)
   P. luteo-virens (Fr.) Maire
   conidial states: *Sepedonium, Acremonium*
   8 species.

. Hypomyces (Fr.) Tul. - Ann. Sci. nat., Bot., Sér. 4, *13*: 11 (1860)
   H. lactifluorum (Schw.) Tul.
   = *Apiocrea* Syd. - Ann. mycol. *18*: 186 (1920)
      A. chrysosperma (Tul.) Syd. = *H. chrysospermus* Tul.
   conidial states: *Cladobotryum, Mycogone, Sepedonium, Sibirina, Trichothecium,*
   Verticillium
   about 20 species (Fig. 67)
   References: Hanlin, 1963, 1964; Gams and Hoozemans, 1970; Arnold, 1969,
   1971.

. Arachnocrea Moravec - Bull. Soc. mycol. Fr. *72*: 161 (1956)
   A. papyracea (Ellis & Holway) E. Müller
   2 species
   References: Müller and von Arx, 1962.

. Gaeumannomyces v. Arx & Olivier - Trans Br. mycol. Soc. *35*: 32 (1952)
   G. graminis (Sacc.) v. Arx & Olivier (= *Ophiobolus graminis* Sacc.) (Fig. 68)
   3 or 4 species, parasitic on stems and roots of cereals and grasses or saprophytic
   References: Schrantz, 1960; Eriksson, 1967; Skou, 1968; Walker, 1972.

. Gnomoniella Sacc. - Michelia *2*: 312 (1881)
   G. tubaeformis (Tode) Sacc.
   related genera are *Sphaerognomonia* Potebnia, *Rehmiella* Wint. and *Heteropera*
   Theiss., cf. von Arx and Müller, 1954.

. Gnomonia Ces. & de Not. - Sferiac. Ital. p. 57 (1863)
   G. setacea (Pers. ex Fr.) Ces. & de Not.
   = *Gnomoniopsis* Berl. - Icones Fung. *1*: 93 (1894)
      G. chamaemori (Fr.) Berl.
   = *Melanopelta* Kirschst. - Ann. mycol. *37*: 113 (1939)
      M. saxonica Kirschst. = *G. sanguisorbae* (Rehm) E. Müller
   = *Rostrocoronophora* Munk - Dansk bot. Arkiv *15*(2): 98 (1953)
      R. geranii Munk = *G. geranii* Hollós
   conidial states: *Cylindrosporella, Discula, Zythia*
   about 50 species (Fig. 69). The genus *Apiognomonia* Höhnel is closely related.

. Cryptodiaporthe Petrak - Ann. mycol. *19*: 118 (1921)
   C. aesculi (Fuckel) Petr.
   conidial states: *Discella, Chondroplea*
   about 10 species, saprophytic or parasitic on woody plants
   References: Müller and von Arx, 1962; Wehmeyer, 1933; Butin, 1957, 1958;
   Kobayashi, 1970.

. Diaporthe Nitschke - Pyrenom. Germ. p. 240 (1870)
   D. eres Nitschke
   synonyms vide Müller and von Arx, 1962; Kobayashi, 1970.

conidial states: *Phomopsis, Libertella*
about 80 species, saprophytic or parasitic on higher plants
References: Wehmeyer, 1933; Kobayashi, 1970.

73. **Endothia** Fr. - Summa Veg. Scand. p. 385 (1849)
   E. gyrosa (Schw.) Fr.
   conidial state: *Endothiella* Sacc. (= *Calopactis* Syd.)
   about 10 species, saprophytic or parasitic. *E. parasitica* (Murr.) Anderson parasitizes *Castanea*
   References: Müller and von Arx, 1962; Kobayashi and Ito, 1956.

74. **Valsa** Fr. - Summa Veg. Scand. p. 410 (1849)
   V. ambiens (Pers. ex Fr.) Fr.
   conidial state: *Cytospora*
   about 20 species
   References: Kobayashi, 1970.

75. **Leucostoma** (Nits.) Höhnel - Ber. deut. bot. Ges. *35*: 631 (1917)
   L. massarianum (de Not.) Höhnel
   conidial state: *Cytospora (Leucocytospora)*
   5 or 6 species
   References: Kern, 1961; Urban, 1958; Kobayashi, 1970.

76. **Phomatospora** Sacc. - Fungi Ven. *2*: 306 (1874)
   P. berkeleyi Sacc.
   = *Phomatosporopsis* Petrak - Ann. mycol. *23*: 37 (1925)
   P. angelicae (Fuckel) Mouton
   about 12 species.

77. **Endoxyla** Fuckel - Symb. mycol., Nachtr. *1*: 321 (1871)
   E. operculata (Alb. & Schw.) Fuckel
   = *Ceratostomella* Sacc. - Michelia *1*: 370 (1878)
   C. vestita Sacc. = *E. cirrhosa* (Pers. ex Fr.) v. Arx & Müller
   2 or 3 species
   References: von Arx and Müller, 1954.

78. **Hercospora** Fr. - Syst. Orbis Veget. p. 119 (1825)
   H. tiliae (Pers.) Fr.
   conidial state: *Rabenhorstia*
   5 species
   References: Müller and von Arx, 1962.

79. **Melanconis** Tul. - Ann. Sci. Nat., Bot., sér. 4, *5*: 109 (1856)
   M. stilbostoma (Fr.) Tul.
   conidial states: *Melanconium, Coryneum, Stilbospora* and others
   about 25 species
   References: Müller and von Arx, 1962 (for synonyms); Wehmeyer, 1941; Kobayashi, 1970.

. **Physalospora** Niessl - Verh. naturf. Ver. Brünn *14*: 10 (1876)
  P. alpestris Niessl
= *Acanthorhynchus* Shear - Bull. Torrey bot. Club *34*: 314 (1907)
  A. vaccinii Shear = *Ph. vaccinii* (Shear) v. Arx & Müller
= *Benedekiella* Verona & Negru - Rev. Pat. veg. *3,4*: 423 (1964)
  B. oxycocci Verona & Negru = *Ph. vaccinii*
= *Pseudoguignardia* Gutner - Mat. Mikol. Fitopat. Ross. *6*: 311 (1927)
  P. scirpi Gutner = *Physalospora scirpi* (Gutner) v. Arx
= *Pseudophysalospora* Höhnel - Ann. mycol. *16*: 57 (1918)
  P. adeana (Rehm) Höhnel = *Ph. adeana* (Rehm) v. Arx & Müller
= *Trichophysalospora* Lebedeva - Acta Inst. Acad. Sci. Pl. Crypt. *1*: 345 (1933)
  T. saviczii Lebedeva = *Ph. saviczii* (Lebedeva) v. Arx & Müller
conidial state: *Papularia*
about 10 species (most of the species described belong to the genera *Glomerella* or *Botryosphaeria*)
References: von Arx and Müller, 1954.

. **Chaetapiospora** Petrak - Sydowia *1*: 86 (1947)
  Ch. islandica (Joh.) Petrak
second species: *Ch. rhododendri* (Tengwall) v. Arx, known esp. in pure culture (Fig. 70a)
References: Müller and von Arx, 1962.

. **Pseudomassaria** Jacz. - Bull. Herb. Boiss. *2*: 662 (1896)
  P. chondrospora (Ces.) Jacz.
synonyms vide Müller and von Arx, 1962; Barr, 1964
12 species; all are saprophytic on twigs and leaves (Fig. 70b).

. **Apiospora** Sacc. - Consp. Gen. Pyren. p. 9 (1875)
  A. montagnei Sacc.
= *Khuskia* Hudson - Trans. Br. mycol. Soc. *46*: 355 (1963)
  K. oryzae Hudson = *A. oryzae* (Hudson) v. Arx comb. nov.
= *Rhabdostroma* Syd. - Ann. mycol. *14*: 362 (1916)
  R. rottboelliae (Rehm) Syd. = *A. montagnei*
= *Scirrhiella* Speg. - Fungi Guar. *1*: 110 (1883)
  S. curvispora Speg. = *A. curvispora* (Speg.) Rehm
conidial state: *Arthrinium, Nigrospora*
4 or 5 species
References: Müller and von Arx, 1962.

. **Leiosphaerella** Höhnel - Sber. K. Akad. Wiss. Wien, math.-nat. Kl. *128*(1): 579 (1919)
  L. praeclara (Rehm) Höhnel
5 species.

. **Ceriospora** Niessl - Verh. naturf. Ver. Brünn *14*: 169 (1876)
  C. dubyi Niessl
= *Microcyclosphaeria* Batista - Rev. Biol. *1*: 301 (1959)
  M. palmicola Batista & Maia = *C. palmicola* (Batista & Maia) E. Müller
5 species
References: Müller and von Arx, 1962.

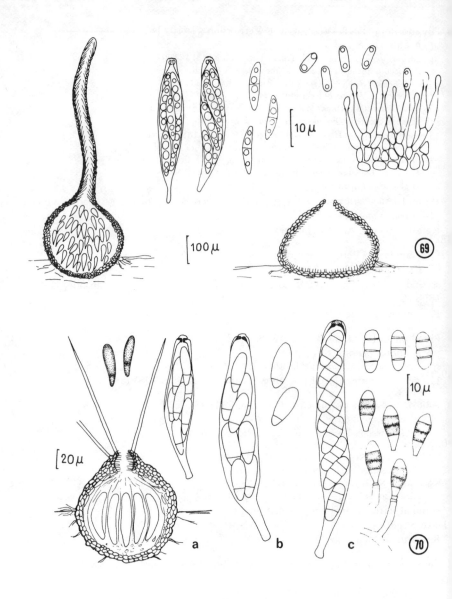

Fig. 69. *Gnomonia fructicola* (Arn.) Fall, perithecium, asci, ascospores and pycnidium, phialides and conidia of the *Zythia* conidial state (orig.).

Fig. 70. a, *Chaetapiospora rhododendri*, perithecium, ascus and ascospores; b, *Pseudomassaria sepincolaeformis* (de Not.) v. Arx, ascus and ascospores; c, *Griphosphaeria corticola*, ascus, ascospores and *Seimatosporium* conidial state (orig.).

86. **Griphosphaeria** Höhnel - Ann. mycol. *16*: 87 (1918)
     G. corticola (Fuckel) Höhnel (Fig. 70c)
     conidial states: *Seimatosporium, Fusarium*

130

second species: *G. nivalis* (Schaffnit) Müller & v. Arx with the conidial state *Fusarium nivale* (Fr.) Ces.
References: Shoemaker and Müller, 1964; Müller and von Arx, 1955, 1973.

7. **Discostroma** Clements - Gen. Fungi p. 50 (1909)
   D. rehmii (Schnabl) Clements = *Metasphaeria massarina* Sacc. (Atti Ist. Venet.
   Sci. Venezia 6,2: 456, 1884) = *Discostroma massarina* (Sacc.) v. Arx comb.
   nov.
   ≡ *Clathridium* Berlese (1897) non Sacc. (1883, 1895)
   ≡ *Curreyella* (Sacc.) Lindau (1897) non Massee (1895)
   = *Phragmodothella* Theiss. & Syd. - Ann. mycol. *13*: 343 (1915)
       P. kelseyi (Ellis & Ev.) Theiss. & Syd. = *D. massarina*
   conidial state: *Seimatosporium*
   References: Shoemaker and Müller, 1964 (as *Clathridium*).

8. **Lepteutypa** Petrak - Ann. mycol. *21*: 276 (1923)
   L. fuckelii (Nits.) Petrak
   = *Hymenopleella* Munk - Dansk bot. Ark. *15*(2): 89 (1953)
       H. hippophaes (Fabre) Munk = *Sphaeria hippophaës* Sollmann = *Lepteutypa hippophaës*
       (Sollmann) v. Arx
   further species: *Lepteutypa indica* (Punithalingam) v. Arx (= *Massaria indica*
   Punithalingam) and *Lepteutypa cupressi* (Nattrass & al.) Swart
   conidial state: *Hyalotiella, Seiridium*
   References: Shoemaker and Müller, 1965; Swart, 1973.

9. **Hypoxylon** Bull. ex Fr. - Summa Veg. Scand. p. 383 (1849)
   H. fragiforme (Pers. ex Fr.) Kickx
   conidial states: *Nodulisporium, Geniculosporium*
   References: Miller, 1961; Martin, 1968, 1969.

0. **Xylaria** Hill ex Grev. - Fl. Edinb. p. 355 (1824) (nom. cons.)
   X. hypoxylon (L.) Grev.
   = *Xylosphaera* Dumortier - Comm. bot. p. 91 (1822)
   conidial state: *Nodulisporium*-like, but stromatic
   References: Martin, 1970; Dennis (1970, using the name *Xylosphaera*).

1. **Rosellinia** de Not. - Giorn. Bot. Ital. *1*: 334 (1844)
   R. aquila (Fr.) de Not.
   conidial states: *Nodulisporium, Dematophora*.

PSEUDOSPHAERIALES, DOTHIORALES

Only species belonging to a relatively small number of genera of these orders develop ascomata with asci and ascospores in pure culture. Many more species grow only with a vegetative mycelium or produce conidia of very different kind. These may belong to genera of the Sphaeropsidales and Moniliales.

Genera including species known to produce ascomata in pure culture can be identified with the following key:

1. Ascospores 1-celled, asci (4-) 8-spored, often clavate, thick-walled 2

1. Ascospores septate or asci many-spored 3

2. Ascomata discrete, ascospores mostly shorter than 20 μm or with slimy appendages (ascigerous states of *Phyllosticta, Selenophoma* or *Kabatia*) . . . .*Guignardia* (1)

2. Ascomata often stromatic, ascospores mostly longer, without appendages (ascigerous states of *Dothiorella, Botryodiplodia* and related genera) . . . . *Botryosphaeria* (2)

3. Ascospores 2-celled, with germpores 4

3. Ascospores without germpores. 5

4. Ascomata ostiolate . . . . *Trichodelitschia* (3)

4. Ascomata non-ostiolate. . . . .*Phaeotrichum* (4)

5. Ascospores with germslits 6

5. Ascospores without germslits 7

6. Ascospores 2-celled . . . . *Delitschia* (5)

6. Ascospores with 3 or more transverse septa . . . .*Preussia* (6)

7. Ascospores small, ornamented or glabrous, primarily 4-celled, separating early in single cells, asci therefore 32-spored, ascomata non-ostiolate. . . . . *Westerdykella* (7)

7. Ascospores not so 8

8. Ascomata non-ostiolate, ascospores 2-celled, pigmented 9

8. Ascomata usually ostiolate or ascospores not 2-celled. 13

9. Ascomata with a continuous wall composed of flattened cells, asci spherical or nearly so, with a thin wall . . . . *Pseudophaeotrichum* (17)

9. Ascomata with a wall composed of plates of radiating cells 10

10. Ascospores glabrous 11

10. Ascospores ornamented 12

11. Asci obovate, ascospore wall not darker near the septum . . . . *Neotestudina* (14)

11. Asci clavate, stalked, ascospore wall darker near the septum . . .*Argynna* (15)

12. Asci clavate, stalked, ascospores echinulate .... *Lepidosphaeria* (16)

12. Asci obovate, ascospores reticulate .... *Testudina* (13)

13. Ascospores 2-celled, hyaline, greenish or brown, without appendages and without a gelatinous sheath 14

13. Ascospores mostly multicelled or filiform or with a gelatinous sheath or with appendages when 2-celled. 20

14. Ascospores hyaline or nearly so 15

14. Ascospores yellowish, greenish or brown 17

15. Ascomata setose, containing a small number of asci, ascospores wider than 9 μm .... *Macroventuria* (8)

15. Ascomata glabrous, ascospores mostly narrower 16

16. Asci without paraphyses, often in fascicles or diverging, ascospores usually slender, ascomata usually small .... *Mycosphaerella* (10)

16. Asci with paraphyses, mostly parallel, ascospores ovoid, often constricted at the septum, ascomata usually medium sized .... *Didymella* (11)

17. Ascospores yellowish, greenish or olive brownish, cells often unequal, ascomata setose or glabrous (parasitic on higher plants) .... *Venturia* (9) (and related genera of the Venturiaceae)

17. Ascospores brown or dark-brown (mostly saprophytic) 18

18. Ascomata erumpent or superficial, mostly hairy or setose . *Herpotrichia* (19)

18. Ascomata immersed, glabrous 19

19. Ascospores shorter than 30 μm, smooth .... *Didymosphaeria* (20)

19. Ascospores large, thick-walled, with a sheath or ornamented .... *Pteridospora* (21)

20. *Drechslera-* (*Helminthosporium-*) or *Curvularia-* conidial states with dark, thick-walled poroconidia present; ascospores filiform or fusiform with transverse septa only, hyaline 21

20. Dark, thick-walled poroconidia absent, ascospores mostly pigmented 22

21. Ascospores filiform, coiled in the ascus .... *Cochliobolus* (22)

21. Ascospores fusiform .... *Setosphaeria* (23)

22. Ascospores with transverse septa only........................... 23

22. Ascospores with transverse and longitudinal septa 32

23. Ostiolar pore provided with short, periphysis-like bristles............. 24

23. Ostiolar pore without such bristles 26

24. Ascospores hyaline, mostly fusiform.................. *Keissleriella* (12)

24. Ascospores pigmented when ripe or not fusiform 25

25. Ascospores cylindrical or fusiform, often with appendages............. . ... *Nodulosphaeria* (25)

25. Ascospores cylindrical filiform, fasciculate in the ascus . . . . *Ophiobolus* (24)

26. Ascospores with a gelatinous sheath, hyaline . . . . *Massarina* (18)

26. Ascospores without sheath or not hyaline. 27

27. Ascomata erumpent or superficial, hairy or setose, ascospores with a central constriction . . . . *Herpotrichia* (19)

27. Ascomata usually immersed, glabrous 28

28. Wall of the ascomata thin, composed of thin-walled cells 29

28. Wall of the ascomata usually thick, parenchymatous 30

29. Ascospores fusiform or nearly so, yellowish or brownish . *Phaeosphaeria* (26)

29. Ascospores cylindrical, rounded at both ends . . . . . . *Paraphaeosphaeria* (27)

30. Ascospores with dark central and light end-cells, wall of the ascomata composed of thin-walled cells . . . . *Trematosphaeria* (28)

30. Central cells of the ascospores usually not darker than the end-cells, wall of the ascomata composed of dark, thick-walled cells · · · · · · · · · · · · · · · · 31

31. Ascospores yellowish or brownish, rarely hyaline, usually fusiform . . . .
. . . . *Leptosphaeria* (29)

31. Ascospores brown, ellipsoidal, 3-septate, with a central constriction . . . .
. . . . *Melanomma* (30)

32. Superficial hyphae dark, constricted at the septa, composed of roundish cells, ascomata superficial, often stalked or erect (sooty moulds) . . . .
. . . . *Capnodium* (31)

32. Superficial hyphae not constricted at the septa. 33

33. Ascospores hyaline or nearly so, without gelatinous sheath, ascomata small, glabrous, paraphyses absent . . . . *Leptosphaerulina* (33)

33. Ascospores not hyaline or with a gelatinous sheath 34

34. Ascomata small, setose, paraphyses absent. . . . . *Dictyotrichiella* (32)

34. Ascomata usually large, paraphyses usually present 35

35. Ascomata aggregated on or immersed in a stroma . . . . *Cucurbitaria* (37)

35. Ascomata usually discrete 36

36. Ascospores flattened in one plane, with 3 or more vertical septa . . . .
. . . . *Clathrospora* (35)

36. Ascospores not flattened or with a single vertical septum. 37

37. Ascospores large, pale, with a gelatinous sheath, with less than 8 septa . . . . . .
. . . . *Pyrenophora* (34)

37. Ascospores various, mostly with 8 or more septa, brown or golden yellow . . .
. . . . *Pleospora* (36)

List of genera

1. **Guignardia** Viala & Ravaz - Bull. Soc. mycol. Fr. *8*: 63 (1892)
   G. bidwelli (Ellis) Viala & Ravaz
   synonyms vide von Arx and Müller, 1954
   about 70 species (Fig. 71), in general only conidial states are formed in pure culture, belonging to the genera *Phyllosticta, Selenophoma* or *Kabatia* (Müller, 1957; Reusser, 1964; van der Aa, 1973).

2. **Botryosphaeria** Ces. & de Not. (1863) sensu Sacc. - Michelia *1*: 42 (1877)
   B. quercuum (Schw.) Sacc.
   synonyms vide von Arx and Müller, 1954
   conidial states: *Botryodiplodia, Lasiodiplodia, Dothiorella* and related genera about 15 species.

3. **Trichodelitschia** Munk - Dansk. bot. Ark. *15*: 109 (1953)
   T. bisporula (Crouan) Munk (Fig. 72b)
   2 species
   References: Müller and von Arx, 1962.

4. **Phaeotrichum** Cain & Barr - Can. J. Bot. *34*: 676 (1956)
   P. hystricinum Cain & Barr (Fig. 72a)
   second species: *P. circinatum* Cain.

5. **Delitschia** Auersw. - Hedwigia *5*: 49 (1866)
   D. didyma Auersw. (Fig. 72c)
   = *Delitschiella* Sacc. - Syll. Fung. *17*: 688 (1905)
      D. polyspora (Griff.) Sacc. = *Delitschia polyspora* Griff.
   = *Pachyspora* Kirschst. - Verh. bot. Ver. Brandenb. *48*: 49 (1906)
      P. gigantea Kirschst. = *D. geminispora* Sacc. & Flag.
   about 15 species
   References: Müller and von Arx, 1962; Cain, 1934; Munk, 1957.

6. **Preussia** Fuckel - Fungi rhenani, Suppl., Fasc. 3, no. 1750 (1866)
   P. funiculata (Preuss) Fuckel (Fig. 73c)
   = *Perisporium* Corda non Fries
   = *Honoratia* Cif. & al. - Atti Ist. bot. Univ. Lab. crittog. Pavia 5,*20*: 176 (1962)
      H. pisana Cif. & al. = *P. isomera* Cain
   = *Sporormiella* Ellis & Ev. - N. Amer. Pyr. p. 136 (1892)
      S. nigropurpurea Ellis & Everh.
   = *Sporormiopsis* Breton & Faurel - Bull. Soc. mycol. Fr. *80*: 247 (1964)
      S. minima (Auersw.) Breton & Faurel = *P. minima* (Auersw.) v. Arx (Fig. 72d)
   about 50 species; some of them may be ostiolate or non-ostiolate
   References: Cain, 1961; von Arx and Storm, 1967; von Arx, 1973; Ahmed and Cain, 1972, using the name *Sporomiella* for ostiolate forms only.

7. **Westerdykella** Stolk - Trans. Br. mycol. Soc. *38*: 419 (1955)
   W. ornata Stolk (Fig. 73a)
   = *Pycnidiophora* Clum - Mycologia 47: 899 (1955)
      P. dispersa Clum = *W. dispersa* (Clum) Cejp & Milko (Fig. 73b)
   5 species
   References: Cejp and Milko, 1964; von Arx and Storm, 1967; von Arx, 1973; Cain, 1961 (sub *Preussia*).

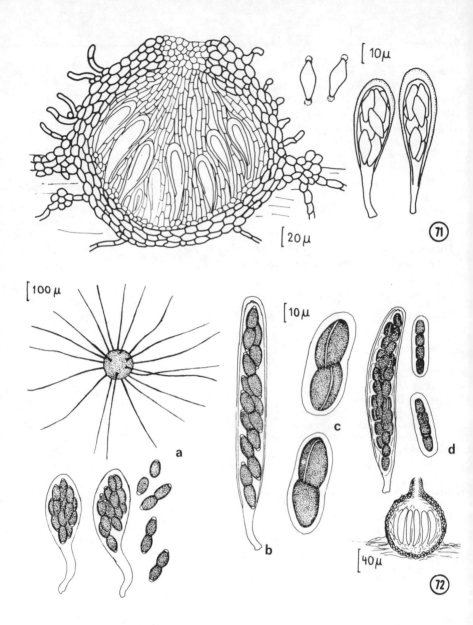

Fig. 71. *Guignardia philoprina*, perithecium, asci and ascospores (orig.).
Fig. 72. a, *Phaeotrichum hystricinum*, ascocarp, asci and ascospores; b, *Trichodelitschia bisporula*, ascus; c, *Delitschia didyma*, ascospores; d, *Preussia minima*, perithecium, ascus and ascospores (a after Cain, 1956; b, c, d orig.).

8. **Macroventuria** van der Aa - Persoonia 6: 359 (1971)
   M. wentii van der Aa
   2 species.

). **Venturia** de Not. sensu Sacc. - Syll. Fung. *1*: 586 (1882)
    V. inaequalis (Cooke) Wint. (cf. Shear, 1948; Korf, 1956)
synonyms vide Müller and von Arx, 1963
conidial states: *Spilocaea, Fusicladium, Pollaccia, Cladosporium*
about 70 species, mostly parasitic on higher plants, esp. on Rosaceae, Salicaceae
and Geraniaceae (Fig. 74)
References: Müller and von Arx, 1962; Menon, 1956; Nüesch, 1960; Barr, 1968.

). **Mycosphaerella** Johanson - Öfvers. Svenska Vet. Akad. *41*: 163 (1884)
    M. punctiformis (Pers. ex Fr.) Schroet. (cf. Wakefield, 1939)
synonyms vide Müller and von Arx, 1962
conidial states: *Cladosporium, Ramularia, Cercospora, Heterosporium, Poly-thrincium, Passalora, Septoria*
Microconidia (pycnidia with spermatia) are classified in the genus *Asteromella*
about 500 species are described, most of them are parasitic on higher plants.
Mycosphaerella tassiana (de Not.) Johanson is the ascigerous state of *Clado-sporium herbarum* (Pers.) Link (Fig. 75)
References: von Arx, 1949; Barr, 1958.

. **Didymella** Sacc. - Syll. Fung. *1*: 545 (1882)
    D. exigua (Niessl) Sacc.
synonyms vide Müller and von Arx, 1962
conidial states: *Ascochyta, Phoma*
about 200 species are described, many are parasitic on higher plants, viz. *D. pinodes* (Berk. & Blox.) Petrak on *Pisum, D. rabiei* (Kov.) v. Arx on *Cicer, D. bryoniae* (Auersw.) Rehm on Cucurbitaceae or *D. cannabis* (Wint.) v. Arx on *Cannabis* (Fig. 76)
References: Corbaz, 1957.

2. **Keisslleriella** Höhnel - Sber. K. Akad. Wiss. Wien, math.-nat. Kl. *128*(1): 592 (1919)
    K. aesculi Höhnel = *K. cladophila* (Niessl) Corbaz
= *Coenosphaeria* Munk - Dansk bot. Ark. *15*(2): 133 (1953)
    C. diaporthoides Munk = *K. cladophila*
= *Trichometasphaeria* Munk - l.c.p. 135 (1953)
    T. dianthi (Rostr.) Munk = *T. gloeospora* (Berk. & Curt.) Holm = *Keissleriella gloeospora* (Berk. & Curt.) v. Arx
= *Zopfinula* Kirschst. - Ann. mycol. *37*: 98 (1939)
    Z. sambucina Kirschst. = *K. cladophila*
References: Bose, 1961.

. **Testudina** Bizz. - Atti Ist. veneto Sci. (Lett. Arti) VI, *3*: 303 (1885)
    T. terrestris Bizz.
= *Marchaliella* Wint. ap. Bommer & Rouss. - Bull. Soc. r. Bot. Belg. *29*: 243 (1891)
    M. zopfielloides Wint. = *T. terrestris*
1 species
References: von Arx, 1971.

. **Neotestudina** Segretain & Destombes - C. r. hbd. Séanc. Acad. Sci. Paris *253*: 2577 (1961)
    N. rosati Segretain & Destombes
1 species.

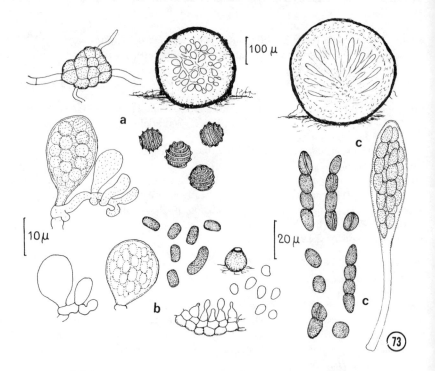

Fig. 73. a, *Westerdykella ornata*, ascocarp initial, ascocarp, asci and ascospores; b, *Westerdykella dispersa*, asci, ascospores and pycnidium, conidiogenous cells and conidia of the *Phoma*-like conidial state; c, *Preussia funiculata*, ascocarp, ascus and ascospores (orig.).

15. **Argynna** Morgan - J. Cincinn. Soc. nat. Hist. *18*: 41 (1895)
    A. polyhedron (Schw.) Morgan
    1 species
    References: Martin, 1941.

16. **Lepidosphaeria** Parguey-Leduc - C. r. hebd. Séanc. Acad. Sci. Paris *270*: 2784 (1970)
    L. nicotiae Parguey-Leduc
    1 species.

17. **Pseudophaeotrichum** Aue & al. - Nova Hedwigia *17*: 84 (1969)
    P. sudanense Aue & al.
    1 species
    References: von Arx, 1971.

18. **Massarina** Sacc. - Syll. Fung. *2*: 153 (1883)
    M. eburnea (Tul.) Sacc.
    synonyms vide Müller and von Arx, 1962

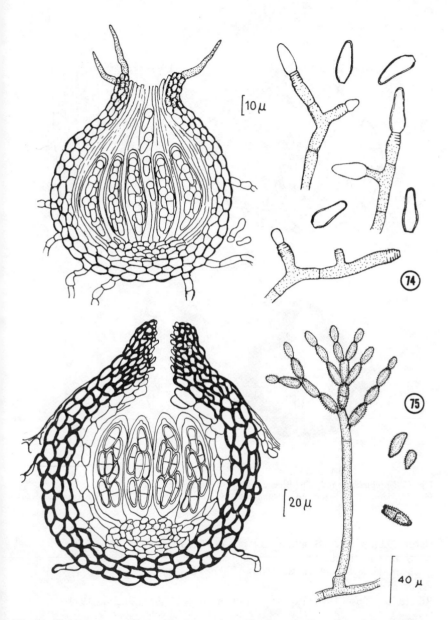

Fig. 74. *Venturia inaequalis*, perithecium and conidiophores and conidia of the *Spilocaea* conidial state (orig.).

Fig. 75. *Mycosphaerella tassiana*, perithecium and conidiophores and conidia of the *Clado-sporium* conidial state (orig.).

about 10 species
References: Holm, 1957; Bose, 1961.

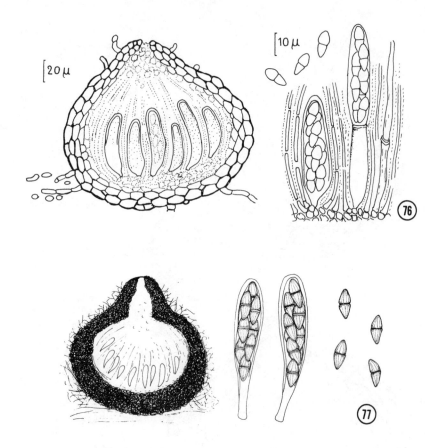

Fig. 76. *Didymella cannabis*, perithecium, asci and ascospores (orig.).
Fig. 77. *Herpotrichia striatispora* Papendorf & v. Arx, perithecium, asci and ascospores (orig.).

19. **Herpotrichia** Fuckel - **Symb.** mycol. p. 142 (1870)
    H. rhenana Fuck.
    synonyms vide Müller and von Arx, 1962
    about 15 species (Fig. 77)
    References: Bose, 1961; Papendorf and von Arx, 1966; Sivanesan, 1971.

20. **Didymosphaeria** Fuckel - Symb. mycol. p. 140 (1870)
    D. futilis (Berk. & Br.) Rehm
    synonyms vide Müller and von Arx, 1962
    about 25 species
    References: Holm, 1957; Scheinpflug, 1958.

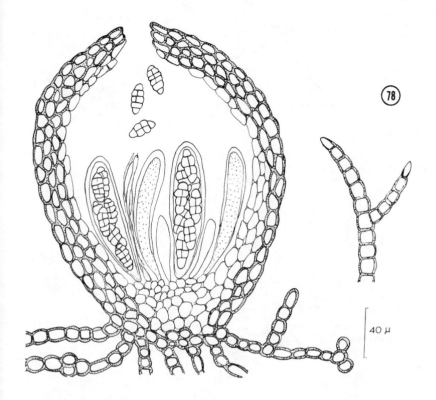

Fig. 78. *Capnodium salicinum*, perithecium with asci and ascospores and mycelium with constrictions (orig.).

1. **Pteridiospora** Penz. & Sacc. - Malpighia *11*: 393 (1897)
   P. javanica Penz. & Sacc.
   3 species
   References: Filer, 1969.

2. **Cochliobolus** Drechsler - Phytopathology *24*: 973 (1934)
   C. heterostrophus (Drechsler) Drechsler
   conidial states: *Helminthosporium* (*Drechslera*), *Curvularia*
   about 10 species, mostly parasitic on cereals and grasses
   References: Shoemaker, 1955; Nelson, 1960, 1964.

3. **Setosphaeria** Leonard & Suggs - Mycologia (in press)
   S. prolata Leonard & Suggs
   further species: *S. turciaca* (Luttrell) Leonard & Suggs, *S. holmii* (Luttrell) Leonard
   & Suggs, all with *Exserohilum* (*Drechslera*)-conidial states.

4. **Ophiobolus** Riess - Hedwigia *1*: 6 (1854)
   O. disseminans Riess = *O. acuminatus* (Sow. ex Fr.) Duby
   synonyms vide Müller, 1952

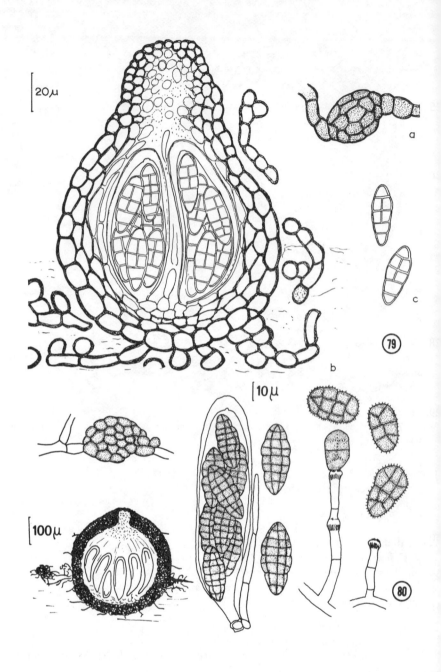

Fig. 79. *Leptosphaerulina australis*, a, perithecial initial; b, ripe perithecium; c, ascospores (orig.).
Fig. 80. *Pleospora herbarum*, perithecial initial, perithecium, ascus, ascospores and the *Stemphylium* conidial state (orig.).

about 100 species. Holm (1957) proposed a new classification of some Pleospora-
ceae and excluded a large number of species and arranged them in different genera
such as *Nodulosphaeria* Rabenh., *Leptospora* Rabenh. or *Entodesmium* Riess.
References: Müller, 1962; Holm, 1948.

25. **Nodulosphaeria** Rabenh. - Herb. mycol. *2*, no. 725 (1858)
     N. derasa (Berk. & Br.) Holm
     about 20 species
     References: Holm, 1957.

26. **Phaeosphaeria** Miyake - J. Coll. Agr. Tokyo *2*: 246 (1910)
     P. oryzae Miyake
     conidial state: *Hendersonia, Septoria*
     about 20 species, mostly on monocotyledonous hosts
     References: Holm, 1957.

27. **Paraphaeosphaeria** O. Eriksson - Ark. Bot. *6*: 405 (1967)
     P. michotii (Westend.) O. Eriksson
     conidial state: *Coniothyrium*
     5 or more species
     References: Hejaroude, 1969.

28. **Trematosphaeria** Fuckel - Symb. mycol. p. 161 (1870)
     T. pertusa Fuckel
     about 12 species
     References: Holm, 1957.

29. **Leptosphaeria** Ces. & de Not. - Sfer. Ital. p. 60 (1861)
     L. doliolum (Pers. ex Fr.) Ces. & de Not.
     synonyms vide Müller, 1950
     about 200 species
     The species placed in *Leptosphaeria* by Müller (1950) and others were divided by
     Holm (1957) into different genera viz. *Leptosphaeria* s. str., *Melanomma* Nits.,
     Nodulosphaeria Rabenh. and *Phaeosphaeria* Miyake.
     References: Müller, 1950; Munk, 1957; Holm, 1952.

30. **Melanomma** Nits. ex Fuckel - Symb. mycol. p. 159 (1870)
     M. pulvis-pyrius (Pers. ex Fr.) Fuckel
     12 or more species, mostly lignicolous
     References: Holm, 1957.

31. **Capnodium** Mont. - Ann. Sci. nat., Bot., Sér. 3, *11*: 233 (1849)
     C. salicinum Mont. (Fig. 78)
     about 25 species, all are sooty moulds
     References: Batista and Ciferri, 1963.

32. **Dictyotrichiella** Munk - Dansk bot. Ark. *15*(2): 132 (1953)
     D. pulcherrima Munk
     second species: *D. mansonii* Schol-Schwarz with the conidial state *Rhinocladiella*
     mansonii (Castell.) Schol-Schwarz
     References: Schol-Schwarz, 1968.

33. **Leptosphaerulina** McAlpine - Fungus Dis. Aust. p. 103 (1902)
   L. australis McAlpine (= *Pseudoplea gaeumannii* (E. Müller) Wehmeyer)
   = *Pseudoplea* Höhnel - Ann. mycol. *16*: 162 (1918)
   P. briosiana (Poll.) Höhnel = *L. briosiana* (Poll.) Graham & Luttrell
   6 species (Fig. 79)
   References: Graham and Luttrell, 1961.

34. **Pyrenophora** Fr. - Summa Veg. Scand. p. 397 (1849)
   P. phaeocomes (Rebent.) Fr.
   = *Macrospora* Fuckel - Symb. mycol. p. 139 (1870)
   Macrospora scirpi Fuckel = *P. scirpicola* (DC. ex Fr.) E. Müller
   conidial state: *Helminthosporium* (*Drechslera*)
   about 10 species
   References: Mü ller, 1951; Drechsler, 1923, 1934; Shoemaker, 1961, 1966.

35. **Clathrospora** Rabenh. - Hedwigia *1*: 115 (1857)
   C. elynae Rabenh.
   about 10 species
   References: Wehmeyer, 1961; Müller, 1951; Simmons, 1952.

36. **Pleospora** Rabenh. - Herb. mycol., edit. 2, p. 347 (1857)
   P. herbarum (Pers. ex Fr.) Rabenh. (Fig. 80)
   synonyms vide Müller, 1951
   about 110 species (many more are described)
   conidial state: *Stemphylium*
   References: Wehmeyer, 1961; Müller, 1951; Simmons, 1952, 1954, 1969; Shoemaker, 1968.

37. **Cucurbitaria** S.F. Gray - Nat. Arr. Brit. Pl. *1*: 519 (1821)
   C. berberidis (Pers.) S.F. Gray
   about 100 species are described
   References: Mirza, 1968.

SPHAEROPSIDALES

16. Not combining above characters 18

17. Phialides arise on branched conidiophores, conidia asymmetrical ....
.... *Chaetomella* (55)

17. Phialides arise basally in the pycnidia on the cells of the pycnidial wall,
conidia cylindrical .... *Amerosporium* (54)

18. Conidia borne pleurogenously on catenulate, cylindrical conidiogenous cells
.... 19

18. Conidia not borne pleurogenously 20

19. Pycnidia setose, discrete......................... *Pyrenochaeta* (11)

19. Pycnidia glabrous, often stromatic .... *Pleurophomella* (46)

20. Conidiogenous cells short, flask-shaped or conical or indistinct 21

20. Conidiogenous cells cylindrical or tapering. 27

21. Conidia rod-shaped, small, not wider than 1.5 μm .... *Asteromella* (3)

21. Conidia not rod-shaped or wider............................... 22

22. Conidia lunate or falcate........................ *Selenophoma* (23)

22. Conidia not lunate or falcate, mostly ellipsoidal or cylindrical 23

23. Pycnidia thick-walled, stromatic, conidia ellipsoidal, slimy, borne from often
dark cells; blastoconidia borne from hyphae or conidia also present ....
.... *Dothichiza* (13)

23. Not combining above characters 24

24. Conidia cylindrical, borne from often dark-walled cells, pycnidia thick-
walled, dark 25

24. Conidia ellipsoidal, ovoidal or short cylindrical, borne from hyaline, flask-
shaped or conical conidiogenous cells 26

25. Pycnidia glabrous, immersed, conidia embedded in a slimy mass ....
.... *Coleophoma* (19)

25. Pycnidia setose, with an irregular wall .... *Chaetopyrena* (20)

26. Conidia usually 2-celled, with a relatively wide, often truncate base, colonies
usually light, but sometimes with dark chlamydospores...... *Ascochyta* (2)

26. Conidia usually 1-celled, with a rounded base, hyphae often dark, sometimes
with 1- or many-celled chlamydospores, pycnidia usually glabrous, sometimes setose or hairy .... *Phoma* (1)

27. Conidia allantoid, small, narrow, borne on filamentous phialides ....
.... *Cytospora* (34)

27. Conidia not allantoid or longer than 10 μm 28

28. Conidiogenous cells (phialides) borne at the base of the pycnidial cavity on a
convex region, conidia usually navicular or falcate.................. 29

28. Conidiogenous cells usually not limited to a basal region 30

29. Pycnidia hemispherical or dimidiate, conidia hyaline . . . .*Pilidium* (53)

29. Pycnidia nearly spherical, conidia brownish in mass . . . . *Pilidiella* (18)

30. Conidia fusiform or nearly so, usually shorter than 9 μm, usually with polar bubbles 31

30. Conidia not fusiform or longer 32

31. Conidia 1-celled (filamentous pseudoconidia often also present) . . . .
. . . .*Phomopsis* (33)
(Pycnidia bright: cf. *Discula*)

31. Conidia at least partly 2-celled . . . . *Discella* (32)

32. Conidia cylindrical, sometimes with slimy appendages, pycnidia often stromatic or dimidiate, large. . . . . *Ceutospora* (44)

32. Conidia not cylindrical 33

33. Pycnidia often hemispherical, conidia borne on phialides, ellipsoidal or ovoidal (parasitic on *Populus*) . . . . *Chondroplea* (31)

33. Pycnidial cavity usually spherical 34

34. Conidia about 3 X as long as wide, often fusiform 35

34. Conidia broadly rounded at the ends or truncate at base 36

35. Conidia with a delicate wall, with slimy caps, sclerotia present . . . .
. . . . *Macrophomina* (25)

35. Conidia without slimy caps, sclerotia absent(?) . . . . *Dothiorella* (24)

36. Conidia with a delicate wall, shorter than 18 μm 37

36. Conidia with a thick or stout wall, usually longer than 18 μm. 45

37. Conidia with a slimy exosporium, often with a slimy appendage or a slimy cap, conidiogenous cells usually dissolving . . . . *Phyllosticta* (22)
(Conidia brownish when mature: cf. *Cleistophoma*)

37. Conidia without appendages, exosporium not slimy, pycnidia often stromatic, blastoconidia on mycelial hyphae present . . . *Phacidiopycnis* (45)

38. Conidia usually 1-celled, rarely septate, shorter than 7 μm, borne on flask-shaped or conical cells in basipetal succession, pycnidial wall usually composed of hyphal elements . . . . *Coniothyrium* (14)

38. Conidia longer or borne on elongated conidiogenous cells 39

39. Conidiogenous cells (phialides) borne at the base of the pycnidium on a convex region, conidia 1-celled, often with a germslit . . . .*Coniella* (16)
(Conidia navicular or lunate, bright, without slit: cf. *Pilidiella*)

39. Conidiogenous cells usually not limited to the base of the pycnidium. . . . 40

40. Conidia 1-celled, blackish brown, with a gelatinous sheath, borne on short, phialide-like cells. . . . .*Melanophoma* (17)

40. Conidia without gelatinous sheath. 41

57. Pycnidia thick-walled, conidia large . . . . *Camarosporium* (10)
57. Pycnidia thin-walled, conidia medium-sized . . . . *Stigmella* (59)

List of genera

. **Phoma** Sacc. - Michelia 2: 4 (1880) (nom. cons.)
  P. herbarum Westend. (Fig. 81)
  = *Plenodomus* Preuss - Linnaea 24: 145 (1851)
    P. rabenhorstii Preuss = *Phoma lingam* (Tode ex Fr.) Desm.
  = *Leptophoma* Höhnel - Sber. K. Akad. Wiss. Wien, math.-nat. Kl. 124: 73 (1915)
    L. acuta (Hoffm. ex Fr.) Höhnel = *Phoma acuta* (Hoffm. ex Fr.) Fuckel
  = *Bakerophoma* Died. - Ann. mycol. 14: 62 (1916)
    B. sacchari Died.
  = *Sclerophomella* Höhnel - Hedwigia 59: 237 (1918)
    S. complanata (Tode ex Desm.) Höhnel = *Phoma complanata* Tode ex Desm.
  = *Polyopeus* Horne - J. Bot., Lond. 58: 139 (1920)
    P. purpureus Horne = *Phoma macrostoma* Mont.
  = *Deuterophoma* Petri - Boll. Real. Staz. Patol. veg. Firenze N. S. 9: 396 (1929)
    D. tracheiphila Petri
  = *Peyronellaea* Goid. - R. C. Accad. Lincei 1: 455, 658 (1946) ex Togliani - Annali sper. agr.,
    Ser. 2, 6: 93 (1952)
    P. glomerata (Corda) Goid. = *P. glomerata* (Corda) Wollenw. & Hochapfel
  ascigerous state: mostly unknown; for a few species of the section 'Plenodomus'
  Leptosphaeria is mentioned. Other species belong to *Didymella* and are mostly
  parasitic on higher plants.
  About 70 species are known in culture, but several hundred are described as
  herbarium specimens.
  References: Boerema, 1964, 1967, 1970; Boerema and Dorenbosch, 1965, 1970,
  1973; Boerema, Dorenbosch and Leffring, 1965; Boerema, Dorenbosch and van
  Kesteren, 1965, 1968; Boerema and Höweler, 1967; Dorenbosch, 1970; Sutton,
  1964; Wollenweber and Hochapfel, 1936.

. **Ascochyta** Lib. - Pl. crypt. Ard. (Fasc. 1) No. 12 (1830); comp. Saccardo,
  Michelia 1: 161 (1878)
  A. pisi Lib.
  = *Diplodina* auct. non Westend.
  ascigerous state: *Didymella* (often incorrectly described as *Mycosphaerella*)
  about 30 species, mostly on higher plants, esp. on Leguminosae (Fig. 82a) (many
  more species are described as herbarium specimens)
  References: Brewer and Boerema, 1965; Wollenweber and Hochapfel, 1936.

. **Asteromella** Pass. & Thüm. - Mycoth. Univ. no. 1689 (1880)
  A. ovata Pass. & Thüm.
  = *Stictochorella* Höhnel - Verh. zool.-bot. Ges. Wien 68: 117 (1918)
    S. heraclei Höhnel
  = *Plectophoma* Höhnel - Sber. K. Akad. Wiss. Wien, math.-nat. Kl. 116: 639 (1907)
    P. umbelliferarum Höhnel
  = *Stictochorellina* Petrak - Ann. mycol. 20: 337 (1922)
    S. carpatica Petrak = *A. carpatica* (Petr.) Petr.

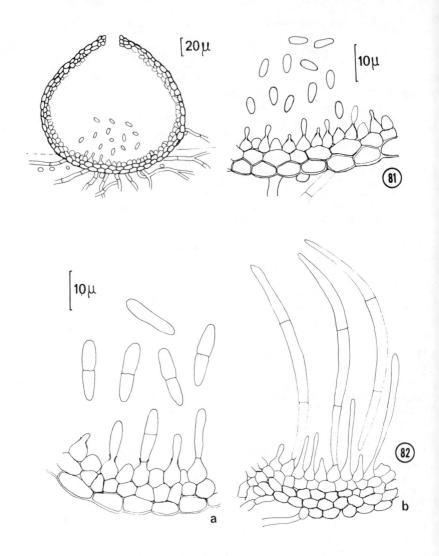

Fig. 81. *Phoma herbarum*, pycnidium and a part of the pycnidial wall with conidiogenous cells and conidia (orig.).
Fig. 82. a, *Ascochyta pisi*, b, *Septoria piricola* Desm., parts of the pycnidial wall with conidiogenous cells and conidia (orig.).

about 50 or more species are described, all are microconidial (spermatial) states of *Mycosphaerella*.
References: von Arx, 1949; Rupprecht, 1959.

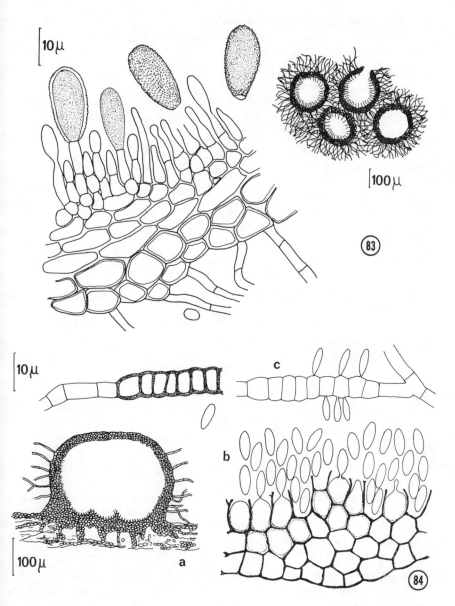

Fig. 83. *Haplosporella acaciae* Petr., group of pycnidia and a part of the pycnidial wall with conidiogenous cells and conidia (orig.).
Fig. 84. *Dothichiza pityophila* (Corda) Petrak, a, pycnidium; b, part of the pycnidial wall with conidiogenous cells and conidia; c, conidiogenous hyphae (orig.).

. **Septoria** Fr. - Syst. mycol. *3*: 480 (1832)
    S. ulmi Fr.
≡ *Phloeospora* Wallr. - Fl. crypt. Germ. *2*: 176 (1833)
    Ph. ulmi (Fr.) Wallr. = *S. ulmi*

ascigerous state: *Mycosphaerella, Sphaerulina*
several hundred species causing leaf spots are described (Fig. 82b)
References: Grove, 1935; Jørstad, 1967.

5. **Stagonospora** (Sacc.) Sacc. - Syll. Fung. *3*: 445 (1884) (nom. cons.)
    S. paludosa (Sacc. & Speg.) Sacc.
    about 100 species are described, only a few are known in pure culture. Some species with broadened, hyaline, 0-4 septate spores would be better classified in *Ascochyta*.
    References: Jørstad, 1967.

6. **Rhabdospora** (Dur. & Mont. ex Sacc.) Sacc. - Syll. Fung. *3*: 578 (1884) (nom. cons.)
    R. oleandri (Dur. & Mont.) Sacc.
    about 100 species are described, most of them are unknown in pure culture.

7. **Hendersonia** Sacc. - Syll. Fung. *3*: 418 (1884) (non Berk.)
    H. sarmentorum Westend.
    about 250 species, only a few are known in pure culture
    References: Allescher, 1903; Archer, 1926; Grove, 1935.

8. **Wojnowicia** Sacc. - Syll. Fung. *10*: 328 (1892)
    W. hirta (Schroet.) Sacc.
    3 or 4 species, *W. graminis* (McAlp.) Sacc. is the cause of a cereal foot rot.

9. **Dichomera** Cooke - Praec. Hend. p. 24 (1878)
    D. saubinetii (Mont.) Cooke
    about 10 species.

10. **Camarosporium** Schulzer - Mykol. Beitr. p. 649 (1870)
    C. quaternatum Schulzer
    = *Thyrococcum* Sacc. - Syll. Fung. *10*: 672 (1892)
      T. punctiforme Sacc.
    about 100 species, only a few are cultivated.

11. **Pyrenochaeta** de Not. - Micr. Ital. *5*: 15 (1845)
    P. nobilis de Not.
    about 50 species, mostly not cultivated. Other species in pure culture do not produce pycnidia. *Pyrenochaeta acicola* (Lév.) Sacc. is *Phoma*-like with setose pycnidia
    ascigerous state: *Herpotrichia, Acanthostigma*
    References: Schneider and Gerlach, 1966.

12. **Haplosporella** Speg. - An. Soc. cient. Argent. *10*: 157 (1880)
    H. parkinsoniae (Speg.) Speg.
    = *Cytosphaera* Diedicke - Ann. mycol. *14*: 205 (1916)
      C. mangiferae Diedicke = *H. mangiferae* (Diedicke) Petrak & Syd.
    = *Pleosphaeropsis* Diedicke - Ann. mycol. *14*: 203 (1916)
      P. dalbergiae Diedicke = *H. dalbergiae* (Diedicke) Petrak & Syd.
    ascigerous state: *Otthia*
    about 80 species, but most of them cannot be distinguished by morphological characters (Fig. 83)
    References: Petrak and Sydow, 1927.

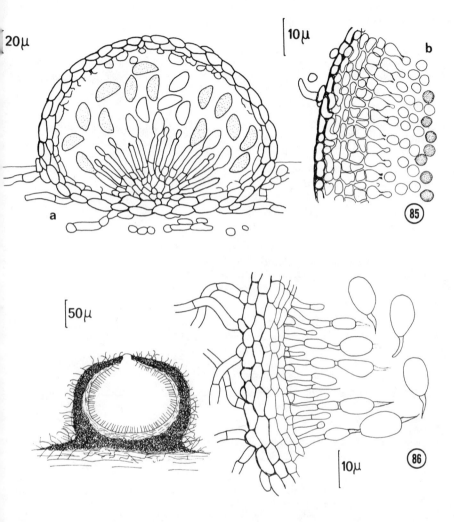

Fig. 85. a, *Coniella diplodiella* (Speg.) Petr. & Syd., pycnidium; b, *Coniothyrium fuckelii* Sacc., part of the pycnidial wall with conidiogenous cells and conidia (orig.).
Fig. 86. *Phyllosticta ilicicola* (Cooke & Ellis) Ellis & Everh., pycnidium and part of the pycnidial wall with conidiogenous cells and conidia (orig.).

5. **Dothichiza** Lib. in Roumeg. - Fungi sel. Gallici exsicc. no. 627 (1880)
     D. sorbi Lib.
= *Blastophoma* Klebahn - Phytopath. Z. *6*: 273 (1933)
     B. thuemeniana Klebahn
= *Dothiopsis* Karst. - Acta Soc. Faun. Fl. Fenn. *6*: 15 (1890)
     D. pyrenophora Karst. = *Dothichiza sorbi*
= *Hormonema* Lagerb. & Melin - Svenska Skogvårdsför. Tidskr. *25*: 219 (1927)
     H. dematioides Lagerb. & Melin = *D. pityophila* (Corda) Petrak

= *Sclerophoma* Höhnel - Sber. K. Akad. Wiss. Wien, math.-nat. Kl. *118*: 1234 (1903)
   S. endogenospora (Sacc.) Höhnel
= *Tylophoma* Klebahn - Phytopath. Z. *6*: 284 (1933)
   T. sorbi Klebahn = *D. sorbi*
ascigerous states: *Dothiora, Sydowia*
about 15 species; in pure culture most of them produce a dark mycelium with blastospores and are hardly distinguishable from *Aureobasidium pullulans* (de Bary) Arn. (Fig. 84)
References: Petrak, 1956; Butin, 1963.

14. **Coniothyrium** Corda - Icon. Fung. *4*: 38 (1840) (nom. cons.) emend. Sacc. - Michelia *2*: 7 (1880)
   C. fuckelii Sacc. (sensu Petrak & Sydow, 1928)
synonyms vide Petrak and Sydow, 1926, Wollenweber and Hochapfel, 1937
ascigerous state: *Leptosphaeria, Paraphaeosphaeria*
about 40 species, but many more are described as herbarium specimens (Fig. 85b)
References: Archer, 1926; Petrak and Sydow, 1927; Bestagno et al., 1958; Wollenweber and Hochapfel, 1937.

15. **Cytoplea** Bizz. & Sacc. - Fl. veneta critt. *1*: 401 (1885)
   C. arundinacea (Sacc.) Petrak & Syd.
= *Phaeophomopsis* Höhnel - Ber. dt. bot. Ges. *35*: 256 (1917)
   P. hederae (Desm.) Höhnel = *C. hederae* (Desm.) Petrak & Syd.
about 15 species
References: Petrak and Sydow, 1927.

16. **Coniella** Höhnel - Ber. dt. bot. Ges. *36*: 316 (1918)
   C. pulchella Höhnel
= *Cyclodomella* Mathur & al. - Sydowia *13*: 145 (1959)
   C. nigra Mathur & al. = *C. diplodiella*
3 or 4 species; *C. diplodiella* (Speg.) Petrak & Syd. is parasitic on *Vitis vinifera*, another probably undescribed species is a common soil fungus (Fig. 85a)
References: Petrak and Sydow, 1927; Sutton, 1969.

17. **Melanophoma** Papendorf - Trans. Br. mycol. Soc. *50*: 503 (1967)
   M. karroo Papendorf
1 species.

18. **Pilidiella** Petrak & Syd. - Phaeospore Sphaerops. p. 462 (1927)
   P. quercicola (Oud.) Petrak & Syd.
= *Anthasthoopa* Subram. & Ramakr. - Proc. Ind. Acad. Sci. *43*: 174 (1956)
   A. simba Subram. & Ramakr.
3 or 4 species. The genus is related to *Coniella*.

19. **Coleophoma** Höhnel - Sber. K. Akad. Wiss. Wien, math.-nat. Kl. *116*: 637 (1907)
   C. crateriformis (Dur. & Mont.) Höhnel
= *Coleonaema* Höhnel - Mitt. bot. Lab. techn. Hochschule Wien *1*: 95 (1924)
   C. oleae (DC.) Höhnel = *C. oleae* (DC.) Petrak & Syd. (Fig. 90b)
about 15 species
References: Petrak & Sydow, 1927.

20. **Chaetopyrena** Pass. - Erb. Critt. Ital. *2*: 1088 (1881)
   Ch. hesperidum Pass. = *Ch. penicillatum* (Fuck.) Höhn.
References: Petrak, 1925.

154

21. **Cleistophoma** Petrak & Syd. - Phaeospore Sphaerops. p. 294 (1927)
   C. suberis (Prill. & Delacr.) Petrak & Syd.
   second species *C. dryina* (Berk. & Curt.) Petrak & Syd.
22. **Phyllosticta** Pers. ex Desm. - Ann. Sci. nat., Bot., Sér. 3, *8*: 28 (1847) (nom. cons.)
   P. convallariae Pers. ex Fr. = *P. cruenta* (Fr.) Kickx
   = *Caudophoma* Patil & Thirum. - Sydowia *20*: 36 (1968)
      C. ehretiae Patil & Thirum.
   = *Phyllostictina* Syd. - Ann. mycol. *14*: 185 (1916)
      P. murrayae Syd.
   ascigerous state: *Guignardia*
   about 50 species (Fig. 86), some of them not distinguishable on morphological characters. More than 2000 species described in the literature have to be excluded and belong to genera such as *Phoma, Phomopsis, Ascochyta* or *Asteromella*.
23. **Selenophoma** Maire - Bull. Soc. bot. Fr. *53*: 87 (1906)
   S. catananches Maire
   = *Falcispora* Bubák & Serebr. - Hedwigia *52*: 269 (1912)
      F. androssowii Bubák & Serebr.
   = *Lunospora* Frandsen - Meddr. Pl. Pat. Afd. Kgl. Vet. Lantb. København *26*: 70 (1943)
      L. oxyspora (Penz. & Sacc.) Frandsen = *Selenophoma donacis* (Pass.) Sprague & Johnson
   = *Selenophomopsis* Petrak - Ann. mycol. *22*: 182 (1924)
      S. juncea (Mont.) Petrak = *Selenophoma juncea* (Mont.) v. Arx
   ascigerous state: *Guignardia*, known for only a few species
   References: Park and Sprague, 1953; Sprague, 1957; Müller, 1957; Jørstad, 1967.
24. **Dothiorella** Sacc. - Michelia *2*:5 (1880)
   D. gregaria Sacc. (Fig. 87a)
   = *Macrophomopsis* Petrak - Ann. mycol. *22*: 108 (1924)
      M. coronillae (Desm.) Petrak = *D. coronillae* (Desm.) Petrak
   = *Cylindrophoma* (Berl. & Vogl.) Höhnel - Ber. dt. bot. Ges. *35*: 351 (1917)
      Macrophoma rimiseda (Sacc.) Berl. & Vogl. = *D. rimicola* (Sacc.) Petrak & Syd.
   ascigerous states: *Botryosphaeria*
   References: Petrak and Sydow, 1927.
25. **Macrophomina** Petrak - Ann. mycol. *21*: 314 (1923)
   M. philippinensis Petrak (= *Dothiorella philippinensis* (Petrak) Petrak)
   second species: *M. phaseoli* (Maubl.) Ashby (= *Dothiorella phaseoli* (Maubl.) Petrak)
   This genus can hardly be distinguished from *Dothiorella*. *M. phaseoli* however is an important root parasite in warmer regions and the name is in use by plant pathologists.
26. **Botryodiplodia** Sacc. - Syll. Fung. *3*: 377 (1884)
   B. juglandicola (Schw.) Sacc. or *B. fraxini* (Lib.) Sacc.
   = *Botryosphaerostroma* Petrak - Hedwigia *62*: 302 (1921)
      B. visci (DC.) Petrak
   ascigerous state: *Botryosphaeria* (= *Physalospora* auct. non Niessl)
   about 25 species, but many more are described
   References: Petrak and Sydow, 1927, 1928; Zambettakis, 1954.

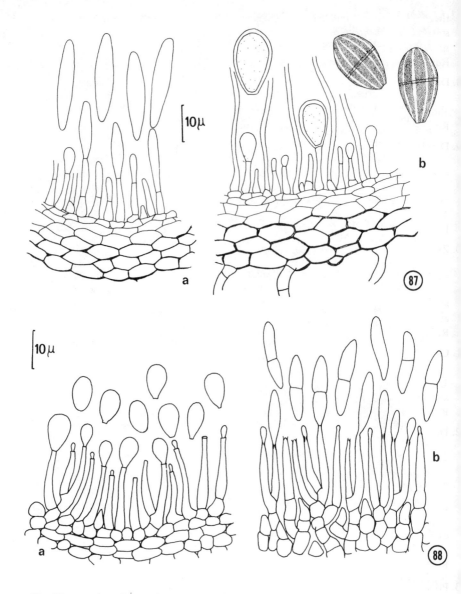

Fig. 87. a, *Dothiorella* conidial state of *Botryosphaeria dothidea* (Moug. ex Fr.) Ces. & de Not.; b, *Lasiodiplodia theobromae*, parts of the pycnidial wall with conidiogenous cells and conidia (orig.).

Fig. 88. a, *Chondroplea populea*; b, *Discella coronata* (Fuck.) Petrak, parts of the pycnidial wall with conidiogenous cells forming conidia in basipetal succession (orig.).

27. **Lasiodiplodia** Griffon & Maubl. - Bull. Soc. mycol. Fr. *25*: 1 (1909)
   L. theobromae (Pat.) Griffon & Maubl. (Fig. 87b)
The fungus is described under a large number of different names. Among these

the oldest may be *Sphaeropsis paradisiaca* Mont. The *Botryosphaeria* ascigerous state is known as *Physalospora rhodina* Berk. & Curt. apud Cooke = *Botryosphaeria rhodina* (Berk. & Curt.) v. Arx
References: Zambettakis, 1954; Goos et al., 1961.

3. **Diplodia** Fr. - Summa Veg. Scand. p. 416 (1849)
   D. conigena Desm. (Ann. Sci. nat., Bot.. Sér. 1, *6*: 69, 1846)
   about 30 species
   References: Zambettakis, 1954.

9. **Darluca** Cast. - Cat. Pl. Marseille, suppl. p. 53 (1845)
   D. filum (Biv.-Bern. ex Fr.) Cast.
   = *Botryella* Syd. - Ann. mycol. *14*: 95 (1916)
   B. nitidula Syd.
   ascigerous state: *Eudarluca*
   4 or 5 species, parasitic on rusts and other fungi
   References: Sydow, 1926; Keener, 1934; Müller and von Arx, 1962.

0. **Zythia** Fr. - Summa Veg. Scand. p. 407 (1849)
   Z. resinae (Ehrenb.) Fr.
   = *Pycnidiella* Höhnel - Sber. K. Akad. Wiss. Wien, math.-nat. Kl. *124*: 91 (1915)
   P. resinae (Ehrenb.) Höhnel = *Z. resinae*
   ascigerous state: *Gnomonia*
   about 20 species (Fig. 69)
   References: Fall, 1951.

1. **Chondroplea** Klebahn - Phytopath. Z. *6*: 229 (1933)
   C. populea (Sacc.) Klebahn (= *Dothichiza populea* Sacc. & Br.) (Fig. 88a)
   1 species
   ascigerous state: *Cryptodiaporthe populea* (Sacc.) Butin
   References: Petrak, 1956; Butin, 1958.

2. **Discella** Berk. & Br. - Ann. Mag. nat. Hist. Ser. 2, *5*: 376 (1850)
   D. carbonacea (Fr.) Berk. & Br. (misapplied) = *D. salicis* (Westend.) Boerema
   = *Cytodiplospora* Oudem. - Contr. Fl. mycol. Pays Bas *15*: 14 (1895)
   C. castaneae Oudem. = *Fusicoccum castaneae* Sacc. = *Discella castaneae* (Sacc.) v. Arx
   = *Diplodina* Westend. - Bull. Acad. Belg. II, *2*: 562 (1857)
   D. salicis Westend. = *Discella salicis* (Westend.) Boerema
   = *Septomyxa* Sacc. - Syll. Fung. *3*: 766 (1884)
   S. aesculi Sacc. = *Discella aesculi* (Corda) Oudem.
   ascigerous state: *Cryptodiaporthe*
   about 10 species (Fig. 88b)
   References: von Arx, 1963; Wehmeyer, 1933; Boerema, 1970.

3. **Phomopsis** Sacc. - Syll. Fung. *18*: 264 (1909) (nom. cons.)
   P. oblonga (Desm.) Höhnel - Sber. K. Akad. Wiss. Wien, math.-nat. Kl. *115*: 680 (1906) (neotype) (conidial state of *Diaporthe eres* Nits.)
   = *Fusicytospora* Gutner - Acta Inst. bot. Acad. Sci. U.S.S.R., Pl. crypt. fasc. *2*: 473 (1935)
   F. mira Gutner
   = *Haplolepis* Syd. - Ann. mycol. *23*: 411 (1925)
   H. polyadelpha Syd.

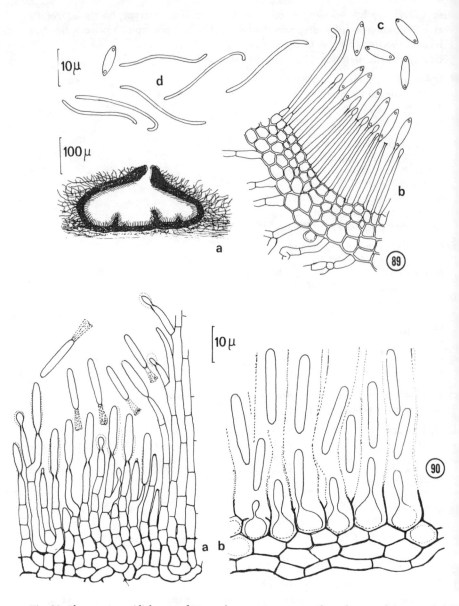

Fig. 89. *Phomopsis* conidial state of *Diaporthe eres* Nits., a, pycnidium; b, part of the pycnidial wall with conidiogenous cells; c, α-conidia; d, β-conidia (orig.).

Fig. 90. a, *Ceutospora phacidioides*; b, *Coleophoma oleae*, parts of the pycnidial wall with conidiogenous cells and conidia (orig.).

= *Myxolibertella* Höhnel - Ann. mycol. *1*: 526 (1903)
    M. aceris Höhnel = *P. aceris* (Höhnel) Sacc.
= *Phomopsella* Höhnel - Mitt. bot. Inst. techn. Hochsch. Wien *5*: 93 (1928)
    P. macilenta (Rob.) Höhnel = *Phomopsis macilenta* (Rob.) Petrak

158

ascigerous state: *Diaporthe*
about 40 species (Fig. 89), but many more are described as herbarium specimens.
References: Wehmeyer, 1933; Grove, 1935; Diedicke, 1915; Wollenweber and
Hochapfel, 1936.

4. **Cytospora** Ehrenb. ex Fr. - Syst. mycol. 2: 540 (1823)
   C. betulina Ehrenb. ex Fr.
   = *Cytonaema* Höhnel - Sber. K. Akad. Wiss. Wien, math.-nat. Kl. *123*: 131 (1914)
      C. spinella (Kalchbr.) Höhnel
   = *Cytophoma* Höhnel - l.c. p. 133
      C. pruinosa (Fr.) Höhnel
   = *Monopycnis* Naumov - Bull. Soc. Oural *35*: 36 (1915)
      M. crataegi Naumov
   ascigerous state: *Valsa* (incl. *Leucostoma*)
   about 40 species, but more than 300 are described
   References: Défago, 1942.

5. **Endothiella** Sacc. - Ann. mycol. *4*: 273 (1906)
   E. gyrosa Sacc.
   ascigerous state: *Endothia*.

6. **Aschersonia** Mont. - Syll. crypt. 260, no. 929 (1856)
   A. taitensis Mont.
   common species: *A. aleyrodis* Webber
   parasitic on scale insects and whiteflies
   ascigerous state: *Hypocrella*
   References: Mains, 1960.

7. **Micropera** Lév. - Ann. Sci. nat., Bot., Sér. 3, *5*: 283 (1846)
   M. drupacearum Lév.
   = *Gelatinosporium* Peck - Rep. N. Y. State Mus. *25*: 48 (1873)
      G. betulinum Peck
   = *Micula* Duby - Hedwigia 2: 8 (1858)
      M. mougeotii Duby
   ascigerous state: *Dermea*
   about 20 species, but morphologically hardly distinguishable.

8. **Brunchorstia** Erikss. - Bot. Zentbl. *47*: 298 (1891)
   B. destruens Erikss. = *B. pinea* (Karst.) Höhnel
   ascigerous state: *Scleroderris lagerbergii* Gremmen = *Crumenula abietina* Lagerberg = *Ascocalyx abietina* (Lagerberg) Schläpfer
   The genus can hardly be distinguished from *Micropera*. *Bothrodiscus* Shear may
   be identical.

9. **Topospora** Fr. - Flor. Scan. Upsal. (1835)
   T. tuberiformis (Fr.) Fr.
   = *Mastomyces* Mont. - Ann. Sci. nat., sér. 3,*10*: 135 (1848)
      M. friesii Mont. = *T. tuberiformis*
   ascigerous state: *Godronia*.

0. **Eleutheromyces** Fuckel - Symb. mycol. p. 183 (1870)
   E. subulatus (Tode) Fuckel

1 species, growing on decaying basidiomycetes
References: Seeler, 1943.

41. **Hyalopycnis** Höhnel - Hedwigia *60*: 151 (1918)
    H. vitrea Höhnel = *H. blepharistoma* (Berk.) Seeler
    1 species, growing on agarics
    References: Seeler, 1943.

42. **Sporonema** Desm. - Ann. Sci. nat., Bot., Sér. *3, 8*: 172 (1847)
    S. phacidioides Desm.
    ascigerous state: *Leptotrochila* (= *Fabraea*)
    about 10 species, mostly causing leaf spots
    References: Schüepp, 1959; Limber, 1955.

43. **Phyctaena** Mont. & Desm. - Plant. crypt. no. 1624 (1847)
    P. vagabunda Desm.
    = *Allantozythia* Höhnel - Ann. mycol. *22*: 203 (1924)
    A. alutacea (Sacc.) Höhnel = *P. vagabunda*
    ascigerous state: *Pezicula*
    References: von Arx, 1957, 1970.

44. **Ceuthospora** Fr. em. Grev. - Scott. cryptog. Fl. *5*: 253 (1827)
    C. phacidioides Grev. ex Fr. (Fig. 90a)
    = *Blennoria* Fr. - Syst. mycol. *3*: 480 (1832)
    B. buxi Fr.
    ? = *Neottiospora* Desm. - Ann. Sci. nat., Bot., Sér. 2, *19*: 335 (1845)
    N. caricum Desm. = *N. caricina* (Desm.) Höhnel
    ascigerous state: *Phacidium*
    about 25 species
    References: Terrier, 1942; Luttrell, 1940; Cunnell, 1957.

45. **Phacidiopycnis** Potebnia - Z. Pfl. Krankh., Pfl. Path. Pfl. Schutz *22*: 129 (1912)
    P. malorum Potebnia = *P. melanconioides* (Peck) v. Arx
    ascigerous state: *Potebniamyces* Smerlis (= *Phacidiella* Potebnia non Karst.)
    2 or 3 species
    References: Smerlis, 1962; Hahn, 1957.

46. **Pleurophomella** Höhnel - Sber. K. Akad. Wiss. Wien, math. nat. Kl. *123*: 123 (1914)
    P. eumorpha (Penz. & Sacc.) Höhnel
    ascigerous state: *Tympanis* Tode ex Fr.
    References: Groves, 1932.

47. **Kellermania** Ellis & Everh. - J. Mycol. *1*: 153 (1885)
    K. yuccigena Ellis & Everh.
    5 species.
    References: Sutton, 1968.

48. **Robillarda** Sacc. - Michelia *2*: 8 (1880)
    R. sessilis Sacc. (Fig. 91a)
    about 10 species
    References: Cunnell, 1958; Marasas et al., 1966; Nicot and Rouch, 1965.

9. **Pseudorobillarda** Morelet - Bull. Soc. Sci. nat. Arch. Toulon Var *175*: 5 (1968)
   P. phragmitis (Cunnell) Morelet
   4 species
   References: Morgan-Jones & al., 1972. *Pseudorobillarda* Nag Ray & al. (Can. J. Bot. *50*: 862, 1972) is a later homonym.

0. **Discosia** Lib. - Fl. Crypt. Ard. no. 345 (1849)
   D. artoceras Tode ex Fr. (Fig. 91b)
   = *Leptina* Batista & Peres - Saccardoa *1*: 25 (1960)
   L. eryobotryae Batista & Peres
   2 or 3 species
   References: Guba, 1961.

1. **Hyalotia** Guba - Monogr. Monochaetia a. Pestalotia p. 292 (1961)
   H. laurina (Mont.) Guba
   8 species.

2. **Hyalotiella** Papendorf - Trans. Br. mycol. Soc. *50*: 69 (1967)
   H. transvalensis Papendorf
   = *Hyalotiopsis* Punithalingam - Mycol. Pap. *119*: 12 (1969)
   Hyalotiopsis subramanianii (Agnihothrudu) Punithalingam = *Hyalotiella transvalensis*
   ascigerous state: *Lepteutypa indica* (Punithalingam) v. Arx.

3. **Pilidium** Kunze - Mycol. Hefte *2*: 292 (1823)
   P. acerinum Kunze
   1 or 2 species
   References: von Höhnel, 1915.

4. **Amerosporium** Speg. - An. Soc. cient. Argent. *13*: 20 (1882)
   A. polynematoides Speg.
   = *Phaeopolynema* Speg. - An. Mus. nac. B. Aires *23*: 117 (1912)
   P. argentinensis Speg. = *A. argentinensis* (Speg.) Petrak & Syd.
   Common species: *A. atrum* (Fuckel) Höhnel = *A. caricum* (Lib.) Sacc. (Fig. 92b)
   References: Petrak, 1965.

5. **Chaetomella** Fuckel - Symb. mycol. p. 401 (1870)
   Ch. oblonga Fuck.
   = *Volutellospora* Mathur & Thirum. - Sydowia *18*: 35 (1965)
   V. cinnamomea Thirum. & Mathur = *Ch. cinnamomea* (Thirum. & Mathur) Petrak
   3 or 4 species (Fig. 92a)
   References: Petrak, 1965; Stolk, 1963.

6. **Dinemasporium** Lév. - Ann. Sci. nat., Bot., Sér. 3, *5*: 274 (1846)
   D. graminum Lév.
   = *Dendrophoma* Sacc. - Michelia *2*: 4 (1880)
   D. cytosporoides Sacc. = *Dinemasporium cytosporoides* (Sacc.) Sutton
   about 10 species
   References: Sutton, 1965.

7. **Stauronema** (Sacc.) Syd. & Butler - Ann. mycol. *14*: 217 (1916)
   S. cruciferum (Ellis) Syd. & Butler
   3 species.

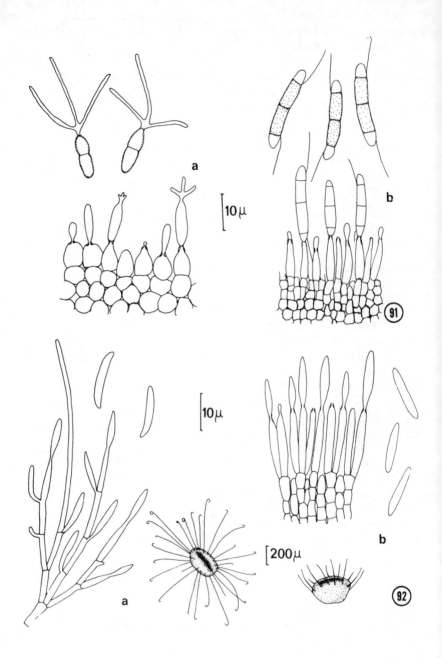

Fig. 91. a, *Robillarda sessilis*; b, *Discosia artoceras*, parts of the pycnidial wall with conidiogenous cells and conidia (orig.).

Fig. 92. a, *Chaetomella circinoseta* Stolk; b, *Amerosporium atrum*, pycnidia, conidiogenous cells and conidia (a from Stolk, 1963, b orig.).

8. **Dilophospora** Desm. - Ann. Sci. nat., Bot., Sér. 2, *14*: 67 (1840)
   D. graminis Desm. = *D. alopecuri* (Fr.) Fr.
   2 or 3 species
   References: Sprague, 1950.

9. **Stigmella** Lév. - Demid. Voy. *2*: 111 (1842)
   S. dryina (Corda) Lév. = *S. effigurata* (Schw.) Hughes

MELANCONIALES

1. Conidiogenous cells rounded above, forming 1-celled conidia by simultaneous budding 2

1. Conidia not formed by simultaneous budding. 3

2. Colonies dark by pigmented hyphae or chlamydospores . . . . . *Kabatiella* (19)

2. Colonies light, often yeast-like (isolated fr. *Quercus, Juglans, Carya* and other trees) . . . . cf. *Microstroma*

3. Conidia hyaline or sometimes light brownish, 1-2(-4)-celled 4

3. Conidia pigmented, mostly dark, truncate at base, often multiseptate 33

4. Colonies restricted in growth, dark by pigmented hyphae or chlamydospores, conidia small, borne in basipetal succession on often firm or pigmented cells (parasitic on plants) 5

4. Not combining above characters 6

5. Conidiogenous cells conical, formed apically in pustulate acervuli composed of vertical rows of cells (isolated fr. conifers) . . . . *Kabatina* (20)

5. Conidiogenous cells cylindrical or tapering, arising from the basal cells of flat acervuli or from small pustules (isolated fr. Vitaceae, Rosaceae and other plants) . . . . *Sphaceloma* (17)

6. Conidia form dark, slimy or dry masses 7

6. Conidia form light, whitish, greyish or reddish masses 8

7. Acervuli small, conidia ellipsoidal, borne on swollen, often curved phialides (isolated fr. *Quercus* and related trees). *Tubakia* (21)

7. Acervuli pustulate, conidia fusiform or cylindrical, borne from cylindrical phialides . . . . cf. *Myrothecium*

8. Acervuli setose or glabrous, phialides firm, often brownish at base, conidia usually longer than 12 μm, usually with granulose content, cylindrical, falcate or ellipsoidal, forming dark appressoria after germination . . . . *Colletotrichum* (3)

8. Not combining above characters 9

9. Basal layer of the acervuli usually dark, pseudoparenchymatous, conidia borne on slender, cylindrical or tapering phialides, often containing apical bubbles (conidial states of Diaporthaceae). 10

9. Basal layer of the acervuli usually light, plectenchymatous, conidia borne on short or elongated conidiogenous cells (often phialides), usually with a truncate base (conidial states of Hypocreaceae, of discomycetes and other ascomycetes) 17

10. Conidia allantoid, acervuli large, stromatic. . . . . *Naemospora* (25)

10. Conidia not allantoid 11

11. Conidia narrow cylindrical or filiform, 1-celled. 12

11. Conidia ellipsoidal, fusiform, falcate or short cylindrical 13

12. Acervuli small, phialides usually cylindrical . . . . *Cylindrosporella* (2)

12. Acervuli large, dark, stromatic, phialides long, tapering . . . . . *Libertella* (26)

13. Conidia 1-celled 14

13. Conidia 2-celled 16

14. Conidia longer than 12 µm, large, acervuli stromatic . . . . *Cryptosporium* (27)

14. Conidia shorter than 12 µm 15

15. Conidia broadly ellipsoidal, phialides shorter than the conidia . . . .
. . . . *Monostichella* (4)

15. Conidia usually narrow, phialides tapering, longer than the conidia . . . .
. . . . *Discula* (23)

16. Conidia falcate (isolated from *Juglans*) . . . . *Marssoniella* (24)

16. Conidia usually straight . . . . cf. *Discella*

17. Conidia spherical or obovate, borne on in the upper part swollen phialides, colonies reddish by an exudate (isolated fr. rust fungi) . . . . *Tuberculina* (18)

17. Not combining above characters 18

18. Conidia ellipsoidal, shorter than 6 µm, conidiogenous cells form dense layers, colonies white (isolated fr. Gramineae) . . . . *Sphacelia* (16)

18. Conidia usually longer 19

19. Colonies orange or salmon-coloured, acervuli stromatic, large, often pulvinate, conidia falcate (isolated fr. insects) . . . . cf. *Aschersonia*

19. Not combining above characters 20

20. Conidia and conidiogenous cells cylindrical or filiform, conidia often in rows, acervuli flat, small 21

20. Not combining above characters 22

21. Acervuli white, conidia cylindrical, often catenulate . . . . *Cylindrosporium* (1)

21. Acervuli not white, conidia filiform or narrow cylindrical . . . .
. . . . *Cylindrosporella* (2)

22. Conidiogenous cells elongated, borne on short conidiophores or laterally on conidiogenous cells, acervuli often pustulate or stromatic 23

22. Conidiogenous cells conical or short cylindrical, arising on the cells of the basal layer, acervuli usually small and flat 25

23. Conidia vermiform or elongated fusiform-falcate . . . . cf. *Phlyctaena*

23. Conidia ellipsoidal, fusiform or cylindrical, up to 4 times as long as broad. 24

24. Acervuli flat, usually small (isolated fr. leaves) . . . . *Cryptocline* (14)

24. Acervuli large, pustulate or stromatic (isolated fr. twigs, fruits, etc.) . . . .
. . . . *Cryptosporiopsis* (15)

25. Conidia 4-celled, cells cruciate, lobed and ciliate ..... *Entomosporium* (12)

25. Cells of the conidia not cruciate.................................... 26

26. Conidia ciliate at both ends, mostly 2-celled, conidiogenous cells often united in a dark, disc-like body .... *Mycoleptodiscus* (13)

26. Conidia not ciliate 27

27. Conidia 1-celled 28

27. Conidia septate 29

28. Conidia rounded above, symmetrical .... cf. *Monostichella* (4)

28. Conidia asymmetrical, falcate or beaked above .... *Gloeosporidiella* (6) (isolated from Gramineae: *Rhynchosporina* (7))

29. Conidia 2-celled 30

29. Conidia not 2-celled 31

30. Conidia rounded above, symmetrical .... *Actinonema* (5)

30. Conidia asymmetrical, usually beaked above .... *Marssonina* (8) (isolated from Gramineae: *Rhynchosporium* (9))

31. Conidia ellipsoidal or broadly cylindrical, with 2 or more septa .... *Septogloeum* (10)

31. Conidia filiform or vermiform 32

32. Acervuli flat, light .... *Phloeosporella* (11)

32. Acervuli cup-shaped, brown .... cf. *Septoria*

33. Conidia with transverse septa only or 1-celled 34

33. Conidia with transverse and longitudinal septa or staurosporous 38

34. Conidia 1-celled borne in basipetal succession, acervuli large, stromatic...... .... *Melanconium* (22)

34. Conidia septate 35

35. Conidia large, thick-walled, borne singly at the end of cylindrical cells...... .... *Coryneum* (33)

35. Conidia borne in basipetal succession............................ 36

36. Basal cell of the conidia with a lateral appendage or without any appendages .... *Seimatosporium* (28)

36. Basal cell of the conidia with an endogenously formed appendage at the truncate base, apical appendages present 37

37. Apical appendage simple .... *Monochaetia* (30)

37. Apical appendage branched .... *Pestalotia* (29) (acervuli small, pustulate: cf. *Pleiochaeta*)

38. Conidia dictyosporous 39

List of genera
(General reference: von Arx, 1970)

1. **Cylindrosporium** Grev. - Scott. cryptog. Fl. *1*: 27 (1823)
   C. concentricum Grev.
   1 species, parasitic on *Brassica*
   References: Thomson, 1936.

2. **Cylindrosporella** Höhnel - Sbr. K. Akad. Wiss. Wien, math.-nat. Kl. *125*: 96 (1916)
   C. carpini (Lib.) Höhnel
   = *Actinonemella* Höhnel in Falck - Mycol. Unters. *1*: 301 (1923)
   A. padi (DC.) Höhnel = *C. padi* (DC.) v. Arx
   = *Gloeosporina* Höhnel - Sbr. etc. *125*: 94 (1916)
   G. inconspicua (Cav.) Höhnel = *C. inconspicua* (Cav.) v. Arx
   = *Pseudothyrium* Höhnel - Mitt. bot. Inst. techn. Hochsch. Wien *4*: 109 (1927)
   P. polygonati (Tassi) Höhnel = *C. polygonati* (Tassi) v. Arx
   ascigerous state: *Gnomonia*
   about 12 species
   References: von Arx, 1957, 1970.

3. **Colletotrichum** Corda in Sturm Deutschl. Fl. Pilze *3*: 41 (1831)
   C. dematium (Fr.) Grove
   = *Colletostroma* Petrak - Sydowia 7: 346 (1935)
   C. baumgartneri Petrak
   = *Colletotrichopsis* Bubák - Öst. bot. Z. *54*: 184 (1904)
   C. piri (Noack) Bubák = *C. gloeosporioides* Penz.
   = *Dicladium* Ces. - Flora *35*: 398 (1852)
   D. graminicola Ces. = *C. graminicola* (Ces.) Wilson
   = *Ellisiella* Sacc. - Michelia *2*: 147 (1880)
   E. caudata Sacc. = *C. graminicola* (Ces.) Wilson
   = *Fellneria* Fuckel - Fungi rhenani no. 1923 (1866)
   F. grossulariae Fuckel = *C. gloeosporioides* Penz.
   = *Gloeosporiopsis* Speg. - An. Mus. nac. Hist. nat. Buenos Aires *20*: 404 (1911)
   G. vinal Speg. = *C. gloeosporioides* Penz.
   = *Steirochaete* Braun & Casp. - Krankh. Pfl. p. 28 (1890)
   S. malvarum Braun & Casp. = *C. malvarum* (Braun & Casp.) Southw.
   = *Vermicularia* Fr. - Summa Veg. Scand. p. 419 (1849)
   V. dematium Fr. = *C. dematium* (Fr.) Grove

167

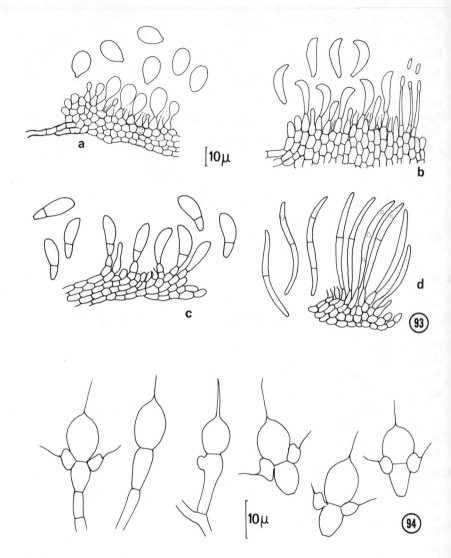

Fig. 93. a, *Monostichella robergei*; b, *Gloeosporidiella ribis*; c, *Marssonina castagnei*; d, *Phloeosporella padi*, parts of acervuli with conidiogenous cells and conidia (orig.).
Fig. 94. *Entomosporium mespili*, conidia (orig.).

ascigerous state: *Glomerella*
about 15 species (Fig. 65)
References: von Arx, 1957, 1970; Simmonds, 1965; Sutton, 1962.

4. **Monostichella** Höhnel - Sber. K. Akad. Wiss. Wien, math.-nat. Kl. *125*: 95 (1916)
   M. robergei (Desm.) Höhnel (Fig. 93a)
   = *Antimanopsis* Petrak - Sydowia 2: 46 (1948)
   A. aequatoriensis Petrak = *M. aequatoriensis* (Petrak) v. Arx

ascigerous state: *Sphaerognomonia, Gnomonia*
4 or 5 species
References: von Arx, 1970.

. **Actinonema** Fr. - Summa Veg. Scand. p. 424 (1849) sensu Grove (1937)
  A. rosae (Lib.) Fr.
ascigerous state: *Diplocarpon*
3 species
References: Klebahn, 1918; Grove, 1937.

. **Gloeosporidiella** Petrak - Hedwigia *62*: 18 (1921)
  G. ribis (Lib.) Petrak (Fig. 93b)
ascigerous state: *Drepanopeziza*
5 species
References: von Arx, 1970; Müller et al., 1958; Rimpau, 1961.

. **Rhynchosporina** v. Arx - Verh. K. Ned. Akad. Wet., afd. Natuurk. *513*: 19 (1957)
  R. meinersii (Sprague) v. Arx, parasitic on grasses.

. **Marssonina** Magnus - Hedwigia *45*: 88 (1906) (nom. cons.)
  M. castagnei (Desm. & Mont.) Magnus (Fig. 93c)
  ≡ *Gloeosporium* Desm. & Mont. - Ann. Sci. nat., Bot., Sér. 3, *12*: 295 (1849)
    G. castagnei Desm. & Mont. = *M. castagnei*
  ≡ *Marssonia* J. C. Fischer (1874) non Karsten (1861)
ascigerous state: *Diplocarpon, Drepanopeziza*
about 60 species, mostly causing leaf spots
References: Grove, 1937; Klebahn, 1918; Rimpau, 1961.

. **Rhynchosporium** Heinsen - Jb. Hamb. wiss. Ver. *18*: 43 (1901)
  R. graminicola Heinsen = *R. secalis* (Oudem.) J.J. Davis
3 species, parasitic on Gramineae
In culture *Rhynchosporium* is indistinguishable from *Marssonina*.
References: Caldwell, 1937; Owen, 1958.

. **Septogloeum** Sacc. - Michelia *2*: 11 (1880)
  S. carthusianum (Sacc.) Sacc.
about 40 species are described, but most of them have to be excluded. The genus
is closely related to *Marssonina* and differs only in the pluriseptate conidia.
References: Petrak, 1953.

. **Phloeosporella** Höhnel - Ann. mycol. *22*: 201 (1924)
  P. ceanothi (Ellis & Ev.) Höhnel
second species: *P. padi* (Lib.) v. Arx (= *Cylindrosporium padi* (Lib.) Karst.) (Fig.
93d)
ascigerous state: *Blumeriella jaapii* (Rehm) v. Arx (= *Coccomyces hiemalis*
Higgins)
References: Blumer, 1958; von Arx, 1961, 1967.

. **Entomosporium** Lév. - Bull. Soc. bot. Fr. *3*: 31 (1856)
  E. maculatum Lév. = *E. mespili* (DC. ex Duby) Sacc. (Fig. 94)
1 species, causing leaf blight on *Cydonia* and *Pyrus*, conidial state of *Diplocarpon
soraueri* (Kleb.) Nannf. (= *Fabraea maculata* Atk.)
References: Klebahn, 1918.

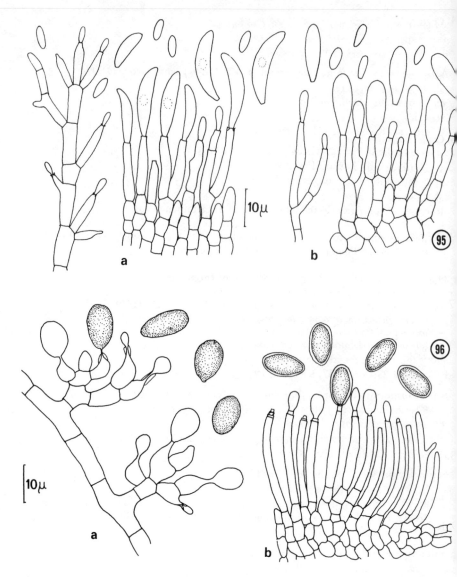

Fig. 95. a, *Cryptosporiopsis malicorticis* (Cordley) Nannf.; b, *Cryptocline cyclaminis* (Sibilia) v. Arx, conidiogenous cells and conidia (orig.).

Fig. 96. a, *Tubakia dryina*; b, *Melanconium bicolor* Nees, conidiogenous parts and conidia (orig.).

13. Mycoleptodiscus Ostazeski - Mycologia 59: 970 (1967)

    M. terrestris (Gerdemann) Ostazeski

  ≡ *Leptodiscus* Gerdemann - Mycologia 45: 552 (1953) non *Leptodiscus* Hertwig (Flagellate)

    L. terrestris Gerdemann = *M. terrestris*

3 species

References: Papendorf, 1967.

4. **Cryptocline** Petrak - Ann. mycol. *22*: 402 (1924)
   C. effusa Petrak
   = *Gloeotrochila* Petrak - Sydowia *1*: 49 (1947)
     G. paradoxa (de Not.) Petrak = *C. paradoxa* (de Not.) v. Arx
   ascigerous state: *Trochila* Fr.
   about 15 species (Fig. 95b)
   Cryptocline is connected with *Cryptosporiopsis* by intermediate forms.
   References: von Arx, 1970; Morgan-Jones, 1973.

5. **Cryptosporiopsis** Bubák & Kabat - Hedwigia *52*: 360 (1912)
   C. scutellata (Otth) Petr.
   synonyms vide von Arx, 1957
   ascigerous states: *Pezicula* (= *Neofabraea*), *Ocellaria*
   about 20 species (Fig. 95a)
   References: von Arx, 1970.

6. **Sphacelia** Lév. - Mém. Soc. Linn. Paris *5*: 578 (1827)
   S. segetum Lév.
   ascigerous states: *Claviceps* Tul., *Epichloë* Fr.
   4 or 5 species, parasitic on graminaceous plants.

7. **Sphaceloma** de Bary - Ann. Oenologie *4*: 165 (1874)
   S. ampelinum de Bary
   = *Manginia* Viala & Pacottet - C.R. Séanc. hebd. Acad. Sci, Paris *139*: 88 (1904)
     M. ampelina Viala & Pacottet = *S. ampelinum*
   = *Melanobasidium* Maubl. - Bull. Soc. mycol. Fr. *22*: 64 (1906)
     M. mali Maubl. = *S. pirinum* (Pegl.) Jenkins
   = *Melanophora* v. Arx - Verh. K. Ned. Akad. Wet., afd. Natuurk. *51*,3: 43 (1957)
     M. crataegi (Dearn. & Barth.) v. Arx = *S. pirinum* (Pegl.) Jenkins
     M. sorbi (Rostr.) v. Arx represents the same fungus
   ascigerous state: *Elsinoë* Rac.
   References: Jenkins and Bitancourt, 1941, 1956, 1966; von Arx, 1970.

8. **Tuberculina** Sacc. - Michelia *2*: 34 (1880)
   T. persicina (Ditmar) Sacc. (Fig. 97a)
   2 or 3 species, parasitic on rust fungi
   References: Hubert, 1935.

9. **Kabatiella** Bubák - Hedwigia *46*: 297 (1907)
   K. microsticta Bubák (Fig. 97b)
   = *Exobasidiopsis* Karakulin - Not. Syst. Inst. crypt. Hort. bot. Petropolis *2*: 101 (1923)
     E. viciae Karakulin = *K. nigricans* (Atk. & Edg.) Karakulin
   = *Pachybasidiella* Bubák & Syd. - Ann. mycol. *13*: 9 (1915)
     P. polyspora Bubák & Syd. = *K. polyspora* (Bubák & Syd.) Karakulin
   = *Polyspora* Lafferty - Scient. Proc. R. Dubl. Soc., n.s. *16*: 258 (1921)
     P. lini Lafferty = *K. lini* (Lafferty) Karakulin
   = *Protocoronospora* Atk. & Edg. - J. Mycol. *13*: 185 (1907)
     P. nigricans Atk. & Edg. = *K. nigricans*
   = *Rhabdogloeopsis* Petrak - Ann. mycol. *23*: 52 (1925)
     R. balsameae (Davis) Petrak = *K. balsameae* (Davis) v. Arx
   about 15 species
   References: Vassiljevsky and Karakulin, 1950; von Arx, 1970.

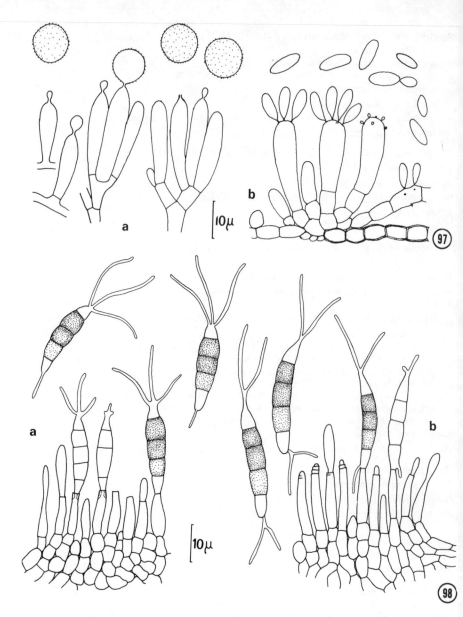

Fig. 97. a, *Tuberculina persicina*; b, *Kabatiella microsticta*, conidiogenous cells and conidia (orig.).
Fig. 98. a, *Pestalotia guepini*; b, *Seimatosporium dilophosporum* (Cooke) Sutton, parts of acervuli with conidiogenous cells and conidia (orig.).

20. **Kabatina** Schneider & v. Arx - Phytopath. Z. 57: 179 (1966)
    K. thujae Schneider & v. Arx
2 species, parasitic on *Thuja* and *Juniperus.*

. **Tubakia** Sutton - Trans. Br. mycol. Soc. *60*: 164 (1973)
    T. japonica (Sacc.) Sutton
≡ *Actinopelte* Sacc. - Ann. mycol. *11*: 315 (1913)
5 species
References: Tehon, 1948; Limber and Cash, 1945; Hering, 1965 (pure culture, as *Cryptocline cinerescens*); Yokoyama and Tubaki, 1971.

. **Melanconium** Link ex Fr. - Syst. mycol. *1*: XL (1821)
    M. atrum Link ex Schlecht.
ascigerous state: *Melanconis* Tul.
about 30 species (Fig. 96b)
References: Kobayashi, 1968; Sutton, 1964; Wehmeyer, 1941.

. **Discula** Sacc. - Syll. Fung. *3*: 674 (1884)
    D. platani (Peck) Sacc.
= *Gloeosporidium* Höhnel - Sber. K. Akad. Wiss. Wien, math.-nat. Kl. *125*: 95 (1916)
    G. platani (Lév.) Höhnel = *D. platani*
ascigerous states: *Apiognomonia, Gnomonia*
about 20 species
References: von Arx, 1957, 1970.

. **Marssoniella** Höhnel - Sber. K. Akad. Wiss. Wien, math.-nat. Kl. *125*: 108 (1916)
    M. juglandis (Lib.) Höhnel
1 species, the conidial state of *Gnomonia leptostyla* Ces. & de Not.
References: Klebahn, 1907.

. **Naemospora** Pers. ex Fr. - Syst. mycol. *3*: 478 (1832) sensu Sacc. - Michelia *2*: 12 (1880)
    N. microspora Desm. (neotypus)
4 or 5 species, conidial state of *Diatrype* Fr.

. **Libertella** Desm. - Ann. Sci. nat., Bot., Sér. 1, *19*: 277 (1830)
    L. betulina Desm.
4 or 5 species, conidial states of *Diatrype, Diatrypella* or *Quaternaria*. A common fungus is *L. faginea* Desm. = *Naemospora crocea* Moug. & Nestl. ex Fr. (!)

. **Cryptosporium** Kunze ex Fr. - Syst. mycol. *3*: 481 (1829)
    C. neesii Corda = *C. vulgare* Fr.
about 20 species, conidial states of *Cryptospora* Tul.

. **Seimatosporium** Corda in Sturm Deutschl. Fl. Pilze, 3, *3*: 13 (1833)
    S. rosae Corda (Fig. 70c)
= *Amphichaeta* McAlpine - Proc. Linn. Soc. N. S. Wales *29*: 118 (1904)
    A. daviesiae McAlpine = *S. daviesiae* (McAlpine) Shoemaker
= *Basipilus* Subram. - Proc. nat. Inst. Sci. India, B, Biol. *27*: 243 (1961)
    B. moravicus (Petr.) Subram. = *S. lichenicola* (Corda) Shoemaker & Müller
= *Cryptostictis* Fuckel - Fungi rhenani no. 1838 (1866)
    C. hysterioides Fuckel = *S. lichenicola*
= *Diseta* Bonar - Mycologia *20*: 299 (1928)
    D. arbuti Bonar = *S. arbuti* (Bonar) Shoemaker
= *Dochmolopha* Cooke - Nuovo G. bot. ital. *10*: 25 (1878)
    D. lonicerae Cooke = *S. lonicerae* (Cooke) Shoemaker
= *Monoceras* Guba - Monogr. Monochaetia Pestalotia p. 290 (1961)

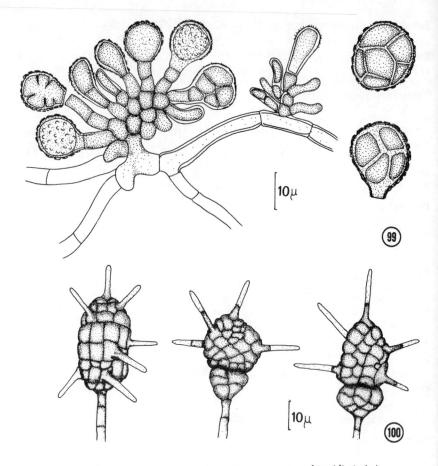

Fig. 99. *Epicoccum purpurascens*, conidiogenous parts and conidia (orig.).
Fig. 100. *Petrakia echinata*, conidia (from van der Aa, 1968).

M. kriegerianum (Bres.) Guba = *S. kriegerianum* (Bres.) Morgan-Jones & Sutton
= *Seiridina* Höhnel - Mitt. bot. Inst. tech. Hochsch. Wien 7: 31 (1930)
    S. rubi Höhnel = *Seimatosporium caudatum* (Preuss) Shoemaker
= *Sporocadus* Corda - Icon. Fung. *3*: 24 (1839)
    S. lichenicola Corda =*Seimatosporium lichenicola*
ascigerous states: *Griphosphaeria, Clathridium*
about 30 species (Fig. 98b)
References: Shoemaker, 1964; Sutton, 1963, 1964.

29. **Pestalotia** de Not. - Mem. R. Accad. Sci. Torino 2, *3*: 80 (1839)
    P. pezizoides de Not.
= *Labridella* Brenckle - Mycologia *22*: 160 (1930)
    L. cornu-cervi Brenckle = *P. cornu-cervi* (Brenckle) Guba
= *Pestalotiopsis* Steyaert - Bull. Jard. bot. État, Bruxelles *19*: 300 (1949)
    P. guepini (Desm.) Steyaert = *Pestalotia guepini* Desm. (Fig. 98a)
= *Truncatella* Steyaert - l.c. p. 293
    T. truncata (Lév.) Steyaert = *P. truncata* Lév.

174

ascigerous state: *Broomella* Sacc., known only for a few species
about 250 species, but often indistinguishable on morphological characters.
References: Steyaert, 1949; Guba, 1961; Dube and Bilgrami, 1966; Shoemaker
and Müller, 1963 (for ascigerous state).

. **Monochaetia** (Sacc.) Allesch. - Rabenh. Krypt. Fl. 1, 7: 665 (1902) (nom. cons.)
  M. monochaeta (Desm.) Allesch.
= *Hyaloceras* Dur. & Mont. - Fl. Alg. Crypt. p. 587 (1846)
  H. notarisii Dur. & Mont. = *M. ceratospora* (de Not.) Guba
= *Seiridium* Nees ex Fr. - Syst. mycol. *3*: 473 (1832)
  S. marginatum Nees ex Fr. = *M. seiridioides* (Sacc.) Allesch.
about 40 species. The genus is closely related to *Pestalotia*
References: Guba, 1961.

. **Petrakia** Syd. - Ann. mycol. *11*: 402 (1913)
  P. echinata (Pegl.) Syd. (Fig. 100)
4 or 5 species
References: Petrak, 1968; van der Aa, 1968; Subramanian, 1957.

. **Epicoccum** Link ex Fr. - Syst. mycol. *3*: 466 (1832)
  E. nigrum Link = *E. purpurascens* Ehrenb. ex Schlecht. (Fig. 99)
= *Cerebella* Ces. - Bot. Ztg. *9*: 669 (1851)
  C. andropogonis Ces. = *E. andropogonis* (Ces.) Schol-Schwarz
2 species
References: Schol-Schwarz, 1959; Ellis, 1971.

. **Coryneum** Nees ex Fr. - Syst. mycol. *3*: 473 (1832)
  C. umbonatum Nees ex Fr.
about 10 species.

. **Asterosporium** Kunze ex Fr. - Syst. mycol. *3*: 484 (1832)
  A. hoffmannii Kunze ex Fr.
2 species.

. **Steganosporium** Corda - Icon. Fung. *3*: 22 (1839)
  S. piriforme (Hoffm.) Corda
about 15 species
References: Moore, 1958.

. **Prosthemium** Kunze ex Fr. - Syst. mycol. *3*: 483 (1832)
  P. betulinum Kunze ex Fr.
second species: *P. stellare* Riess.

MONILIALES

General key to series

Moniliales Key I (Genera with clamp connexions)

1. Aerial conidia violently expelled .... cf. Sporobolomycetales
1. Aerial conidia not violently expelled 2
2. Conidia composed of 4 cruciate cells .... *Riessia* (2)
2. Conidia not cruciate, usually 1-celled........................... 3
3. Conidia borne on sterigmata-like outgrowths of hyphae or conidiophores, thin-walled (parasitic on nematodes) .... *Nematoctonus* (1)
3. Conidia not borne on sterigmata-like outgrowths................... 4
4. Conidia borne simultaneously on a swollen clavate head on denticles ......
   .... *Spiniger* (5)
4. Conidia not borne simultaneously on a swollen head................. 5
5. Colonies partly yeast-like, conidia (basidiospores?) borne in a terminal cluster on erect, thick-walled conidiophores .... cf. *Filobasidium*
5. Not combining above characters............................... 6
6. Conidia usually thick-walled and strikingly pigmented, with a truncate base or truncate at both ends 7
6. Conidia cylindrical, spherical, clavate or ellipsoidal, thin-walled, usually hyaline ....
   conidial states of basidiomycetes, e.g. of *Trametes, Lenzites, Polyporus, Bjerkandera, Serpula, Pholiota, Coprinus* or *Collybia.*
7. Conidia catenulate .... *Ptychogaster* (4)
7. Conidia not catenulate........................... *Sporotrichum* (3)

List of genera

1. **Nematoctonus** Drechsler - Mycologia *38*: 1 (1946)
   N. haptocladus Drechsler
   8 species (Fig. 101b)
   References: Jones, 1964; Cooke and Godfrey, 1964.

2. **Riessia** Fres. - Beitr. Mykol.: 74 (1852)
   R. semiophora Fres. (Fig. 101c)
   1 species
   References: Goos, 1967.

3. **Sporotrichum** Link ex Fr. - Syst. mycol. *3*: 415 (1832)
   S. aureum Link (Fig. 101d)
   4 or 5 species. A large number of other species described as *Sporotrichum* have to be excluded and may belong to genera such as *Sporothrix, Chrysosporium, Rhinocladiella* or *Beauveria* (compare Müller, 1964, 1965).
   References: von Arx 1971, 1973.

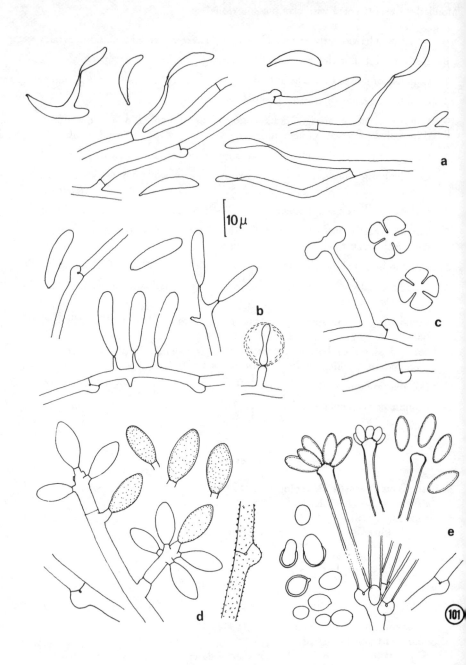

Fig. 101. a, *Itersonilia* spec.; b, *Nematoctonus concurrens* Drechsler; c, *Riessia semiophora*; d, *Sporotrichum aureum*; e, *Filobasidium capsuligenum* (Fell & al.) Rodrigues de Miranda (a from Moreau, 1963, b from Drechsler, 1946, c from Goos, 1967, d, e orig.).

4. **Ptychogaster** Corda - Icones Fungorum 2: 24 (1838)
   P. albus Corda = *P. fuliginoides* (Pers. ex Steudel) Donk, conidial state of
   Tyromyces ptychogaster (Ludwig) Donk.
   Further species: *Ptychogaster rubescens* Boud. and *P. citrinus* Boudier
   References: von Arx 1973.

5. **Spiniger** Stalpers (in press)
   S. meineckellus (Olson) Stalpers
   conidial states of *Hyphoderma, Heterobasidion* and other basidiomycetes
   References: McKeen, 1952 (as *Peniophora*).

Moniliales Key II (genera with arthroconidia)

1. Arthroconidia (fission cells, thalloconidia) and blastoconidia are formed .. 2
1. Blastoconidia absent (chlamydospores or endoconidia may be present). ... 4
2. Blastoconidia borne simultaneously on swollen heads *Trichosporonoides* (10)
2. Blastoconidia not borne simultaneously on swollen heads 3
3. Blastoconidia borne in acropetal chains (osmophilic). .... *Moniliella* (1)
3. Blastoconidia usually not in acropetal chains, colonies yeast-like ....
   .... cf. *Trichosporon*
4. Endoconidia borne in hyphal cells or in arthroconidia, colonies flat, spreading, yeast-like .... cf. *Protendomycopsis*
4. Not combining above characters. 5
5. Fertile hyphae aggregated in synnemata, conidia borne inside the fertile hyphae .... *Briosia* (11)
5. Synnemata absent 6
6. Colonies orange or reddish, spreading, conidia cylindrical, hyaline, borne in sporodochia, in often branched chains .... *Cylindrocolla* (12)
6. Sporodochia usually absent. 7
7. Fertile hyphae borne on flask-shaped cells, conidia yellowish or reddish, cubic or spherical, colonies restricted, reddish (osmophilic). ... *Wallemia* (3)
7. Not combining above characters 8
8. Fertile hyphae borne on conidiophores 9
8. Conidiophores absent 12
9. Conidiophores wide, apically branched or swollen, conidia barrel-shaped, with separating cells .... cf. *Amblyosporium*
9. Not combining above characters 10
10. Fertile hyphae borne in sympodulae on geniculate conidiophores. ....
    .... *Sympodiella* (9)
10. Fertile hyphae not borne in sympodulae on geniculate conidiophores 11
11. Conidiophores long, pigmented, apically branched ...... *Oidiodendron* (2)
11. Conidiophores short, hyaline .... cf. *Chrysosporium*
12. Dark, often catenulate chlamydospores present .... *Scytalidium* (6)
12. Chlamydospores absent or hyaline 13
13. Conidial chains coiled, often also branched (usually thermophilic) ....
    .... *Malbranchea* (7)
13. Conidial chains not coiled 14
14. Colonies light, mostly whitish, conidia not borne inside the hyphae. 15

List of genera

.. **Moniliella** Stolk & Dakin - Antonie van Leeuwenhoek *32*: 399 (1966)
    M. acetoabutens Stolk & Dakin (Fig. 102a)
second species: *Moniliella suaveolens* (Lindner) v. Arx. Both species are osmophilic.

2. **Oidiodendron** Robak - Nyt Mag. Naturvid. *71*: 243 (1932)
    O. fuscum Robak
= *Stephanosporium* dal Vesco - Allionia 7: 182 (1961)
    S. atrum dal Vesco = *O. cereale* (Thuem.) Barron
In *O. cereale* the conidia are oblate, provided with a dark equatorial girdle and this species therefore can also be classified in its own genus as *Stephanosporium cerealis* (Thüm.) Swart.
about 10 species (Fig. 102b)
References: Barron, 1962; Swart, 1965; Morrall, 1968.

3. **Wallemia** Johan-Olson - Forh. Christiania Vid.-Selsk. *12*: 6 (1887)
    W. ichthyophaga Johan-Olson = *Sporendonema sebi* Fr.
    = *Wallemina sebi* (Fr.) v. Arx (Fig. 102c)
= *Bargellinia* Borzi - Malpighia 2: 469 (1888)
    B. monospora Borzi = *W. sebi*

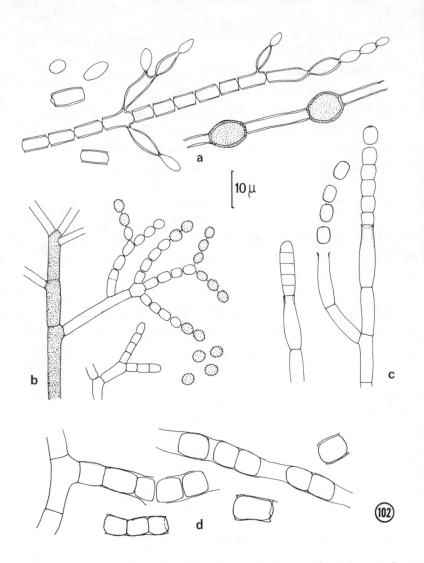

Fig. 102. a, *Moniliella acetoabutens;* b, *Oidiodendron tenuissimum* (Peck) Hughes; c, *Wallemia sebi;* d, *Sporendonema purpurascens* (a from Stolk and Dakin, 1966, b from Barron, 1962, c, d orig.).

= *Hemispora* Vuill. - Bull. Soc. mycol. Fr. *22*: 125 (1906)
   H. stellata Vuill. = *W. sebi*

Wallemia sebi is better known under names such as *Sporendonema epizoum* (Corda) Cif. & Red., *Hemispora stellata* Vuill. or *Sporendonema sebi* Fr.

4. **Sporendonema** Desm. ex Fr. - Syst. mycol. *3*: 434 (1832)
   S. casei Desm. (Fig. 102d)
= *Coprotrichum* Bon. - Handb. allg. Mykol. p. 17 (1851)
   C. purpurascens Bon. = *S. purpurascens* (Bon.) Mason & Hughes

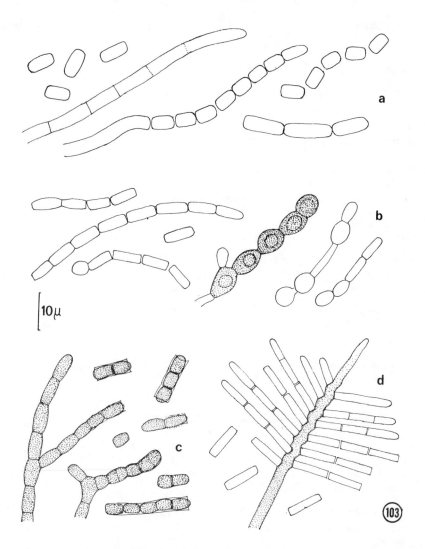

Fig. 103. a, *Geotrichum candidum*; b, *Scytalidium lignicola*; c, *Bahusakala* spec.; d, *Sympodiella acicola* (a, b, c orig., d from Kendrick, 1958).

= *Allonema* Syd. - Ann. mycol. *32*: 283 (1934)
    A. roseum (Grove) Syd. = *S. purpurascens*
2 or 3 species.

5. **Geotrichum** Link ex Pers. - Mycol. europ. *1*: 26 (1822)
    G. candidum Link ex Pers. (Fig. 103a)
= *Oosporoidea* Sumst. - Mycologia *5*: 53 (1913)
    O. lactis (Fres.) Sumst. = *Oospora lactis* (Fres.) Sacc. = *G. candidum*

ascigerous states: *Endomyces, Dipodascus.* Some genera in Eurotiales also have *Geotrichum*-like conidial states.
about 10 species
References: Saëz, 1957; Carmichael, 1957; Morenz, 1963, 1964; von Arx, 1972.

6. **Scytalidium** Pesante - Ann. Sper. Agr., n.s. *11*: 249 (1957)
   S. lignicola Pesante (Fig. 103b)
   3 species
   References: Klingström and Beyer, 1965.

7. **Malbranchea** Sacc. - Michelia *2*: 639 (1882)
   M. pulchella Sacc. & Penzig
   = *Thermoidium* Miehe - Ber. dt. bot. Ges. *35*: 510 (1910)
   T. *sulphureum* Miehe = M. *pulchella* v. *sulphurea* (Miehe) Cooney & Emerson
   3 or 4 species
   References: Baldacci et al., 1938; Cooney and Emerson, 1964.

8. **Bahusakala** Subram. - J. Ind. bot. Soc. *37*: 63 (1958)
   B. olivaceo-nigra (Berk. & Br.) Subram.
   2 or 3 species (Fig. 103c).

9. **Sympodiella** Kendrick - Trans. Br. mycol. Soc. *41*: 519 (1958)
   S. acicola Kendrick (Fig. 103d)
   1 species.

10. **Trichosporonoides** Haskins & Spencer - Can. J. Bot. *45*: 519 (1967)
    T. oedocephalis Haskins & Spencer
    1 species.

11. **Briosia** Cavara - Atti Ist. bot. Pavia *2*(1): 321 (1888)
    B. ampelophaga Cavara
    = *Coremiella* Bubak & Krieger - Ann. mycol. *10*: 52 (1912)
    C. *cystopoides* Bubak & Krieger = B. *cystopoides* (Bubak & Krieger) v. Arx
    4 or 5 species
    References: von Arx, 1972.

12. **Cylindrocolla** Bon. - Handb. allg. Mykol. p. 149 (1851)
    C. urticae (Pers. ex Fr.) Bon.
    1 or 2 species.

Moniliales Key III (Genera with conidia borne in basipetal succession)

1. Conidia borne endogenously within tube-like conidiogenous cells 2
1. Conidia not borne endogenously within cylindrical tubes 5
2. Endoconidia septate . . . . *Sporoschisma* (4)
2. Endoconidia 1-celled 3
3. Pigmented chlamydospores absent . . . . *Chalara* (1)
3. Pigmented chlamydospores present 4
4. Chlamydospores 1-celled, not in chains . . . . *Chalaropsis* (2)
4. Chlamydospores catenulate or with transverse septa . . . . *Thielaviopsis* (3)
5. Conidiogenous cells coarse, brown, apically integrated in erect, septate conidiophores, synnemata or sporodochia absent 6
5. Not combining above characters 8
6. Conidia thick-walled, truncate at base . . . . *Catenularia* (76)
6. Conidia hyaline, thin-walled, not truncate 7
7. Conidia small, hyaline, 1-celled, without appendages . . . . . . *Chloridium* (75)
7. Conidia curved, septate or with appendages . . . . *Codinaea* (72)
8. Thick-walled aleurioconidia (chlamydospores) predominant, phialoconidia
   usually 1-celled or small . . . . see Key V.
8. Aleurioconidia absent or present, but not predominant 9
9. Conidia with a rounded, pointed or papillate base, borne on usually typical
   phialides 10
9. Conidia with a truncate base, separating from the often elongating conidiogenous cell by a septum 82
10. Phialides usually in pairs above each other, the basal one with a lateral tube,
    conidia 1-celled, catenulate, hyaline . . . . *Sesquicillium* (23)
    (Conidiogenous cells in sporodochial tufts: see Melanconiales)
10. Phialides not so 11
11. Phialides usually cylindrical, at least partly intercalary in chains, with a later-
    al opening and a collarette, conidia 1-celled, colonies spreading, light or dark.
    . . . . *Cladorrhinum* (50)
11. Phialides not in chains 12
12. Conidiophores absent or not differentiated, phialides formed on undif-
    ferentiated hyphae or on hyphal strands 13
12. Conidiophores, synnemata or sporodochia present 21
13. Conidia usually 2-celled 14
13. Conidia usually 1-celled 15

185

14. Conidia hyaline, borne obliquely on filamentous conidiogenous cells ......
....*Trichothecium* (80)
14. Conidia brown, borne on often swollen conidiogenous cells .. *Exophiala* (11)
15. Phialides with a median swelling or/and a thick-walled, cylindrical base. ... 16
15. Phialides widest near the base or flask-shaped 17
16. Phialides inflated near the base, conidia spherical, catenulate.............
....*Torulomyces* (8)
16. Phialides with a thick-walled base, conidia not spherical ... *Monocillium* (7)
17. Besides phialoconidia also blastoconidia borne on scars present, colonies dark, often yeast-like, restricted in growth........... *Rhinocladiella* (10)
17. Blastoconidia absent 18
18. Phialides flask-shaped, swollen, often with collarette, mostly brownish, conidia hyaline or brownish .... *Phialophora* (9)
18. Phialides gradually tapering or pointed 19
19. Phialides usually reduced to sterigmata-like pegs forming a single conidium, colonies lanose, white .... *Aphanocladium* (29)
19. Phialides not reduced to sterigmata-like pegs 20
20. Hyphae narrow, conidia usually small, in droplets or in chains ....
.... *Acremonium, Gliomastix* (5, 6)
20. Hyphae wider, phialides often on short branches or in verticils, septate, larger conidia often present .... cf. *Verticillium, Fusarium, Cylindrocarpon*
21. Conidia catenulate, usually forming dry masses 22
21. Conidia not catenulate 40
22. Sporodochia present 23
22. Sporodochia absent, synnemata present or absent................... 25
23. Sporodochia setose, conidia usually cylindrical .... *Volutina* (28)
23. Sporodochia glabrous 24
24. Phialides and conidia cylindrical, in mass greenish ...... *Metarrhizium* (26) (conidial masses dark, usually blackish green: cf. *Myrothecium*)
24. Phialides flask-shaped, conidia ellipsoidal or fusiform ... *Nalanthamala* (37)
25. Conidia septate 26
25. Conidia 1-celled 28
26. Conidia falcate or asymmetrical, pigmented, multiseptate ... *Fusariella* (63)
26. Conidia ellipsoidal or cylindrical, hyaline........................ 27
27. Conidia cylindrical, usually multiseptate .... *Penicillifer* (61)
27. Conidia obovoidal or ellipsoidal, usually 2-celled ..... *Cladobotryum* (81)

28. Phialides widest near the apex, dark, borne in apical clusters, conidia roundish, verrucose, dark . . . . *Memnoniella* (14)

28. Not combining above characters 29

29. Conidiophores with an apical swelling. 30

29. Conidiophores without apical swelling 31

30. Synnemata present, phialides cylindrical, without distinct neck (entomogenous) . . . . *Gibellula* (59)

30. Synnemata absent, phialides flask-shaped or tapering . . . . . . *Aspergillus* (15)

31. Conidiophores stout, dark, thick-walled, apically branched, with prophialides and phialides . . . . *Thysanophora* (54)

31. Conidiophores not so 32

32. Conidia spherical, verrucose, in short chains, conidiophores short *Eladia* (17)

32. Not combining above characters 33

33. Conidia longer than 20 μm, dark, biapiculate. . . . . *Phialomyces* (22)

33. Conidia usually shorter, mostly hyaline. 34

34. Conidiophores usually with an apical penicillus, conidia usually roundish or ellipsoidal, colonies often greenish, synnemata absent or present, chlamydospores mostly absent . . . . *Penicillium* (16)

34. Not combining above characters 35

35. Phialides short, swollen, apically rounded, clustered in whorls (entomogenous) . . . . *Nomuraea* (20)

35. Phialides flask-shaped, usually with a tapering or tube-like neck 36

36. Conidiophores upright, warty, brown, usually simple, conidia smooth or ornamented, often fusiform . . . . *Acrophialophora* (19)

36. Conidiophores light, branched, often forming synnemata 37

37. Conidia fusiform, connected with slimy appendages, phialides cylindrical, apically with a long, narrow tube . . . . *Phialotubus* (25)

37. Not combining above characters. 38

38. Synnemata present, phialides form dense layers (entomogenous) . . . .
. . . . *Akanthomyces* (27)

38. Synnemata absent or present, phialides divergent, not in dense layers 39

39. Conidia obliquely arranged, chains becoming slimy . . . . *Mariannaea* (21)

39. Conidia not obliquely arranged, chains long, dry . . . . *Paecilomyces* (18)

40. Conidiogenous cells arise laterally on erect, seta-like conidiophores. 41

40. Conidiophores not seta-like 42

41. Conidiogenous cells phialidic, with one apical opening. . . . *Chaetopsina* (51)

41. Conidiogenous cells with several conidiogenous openings. . . *Chaetopsis* (52) (conidiogenous cells recurved, with 1 to several denticles: cf. *Zygosporium*)

89. Synnemata hairy or setose . . . . *Trichurus* (79)

89. Synnemata not hairy . . . . *Cephalotrichum* (78)

90. Dark aleurioconidia with a germslit present . . . . cf. *Wardomyces*

90. Dark aleurioconidia absent 91

91. Conidia borne on dichotomously branched (at branching not septate) conidiogenous cells . . . . *Polypaecilum* (87)

91. Conidiogenous cells not branched 92

92. Conidiogenous cells without annellations, conidia wider than 8 μm, hyaline
. . . . *Basipetospora* (88)

92. Conidiogenous cells phialide-like, often with annellations, conidia less than
8 μm wide, usually pigmented . . . . *Scopulariopsis* (77)

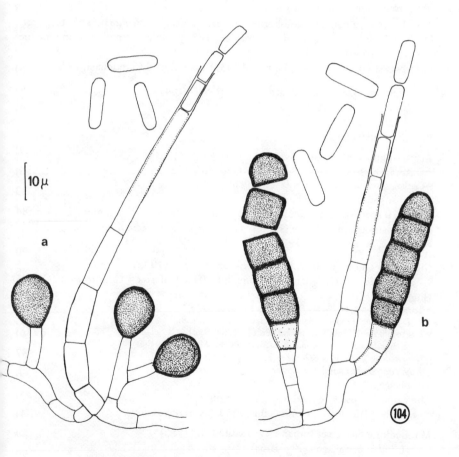

Fig. 104. a, *Chalaropsis thielavioides*; b, *Thielaviopsis basicola* (orig.).

191

List of genera

1. **Chalara** Corda - Icon. Fung. *2*: 9 (1838)
   C. fusarioides Corda
   about 20 species, most of them are known to be conidial states of *Ceratocystis* (Fig. 55)
   References: Hunt, 1956.

2. **Chalaropsis** Peyr. - Boll. Staz. Sper. Agr. Ital. *49*: 595 (1916)
   C. thielavioides Peyr. (Fig. 104a)
   second species: *C. punctulata* Hennebert, and compare *Ceratocystis*
   References: Hennebert, 1967; Hunt, 1956; Sugiyama, 1968.

3. **Thielaviopsis** Went - Meded. Proefst. W-Java 7: 4 (1893)
   T. ethacetica Went = *T. paradoxa* (de Seynes) Höhnel
   = *Hughesiella* Batista & Vital - An. Soc. biol. Pernambuco *14*: 141 (1956)
   H. euricoi Batista & Vital = *T. paradoxa*
   ascigerous state: *Ceratocystis*
   3 or 4 species, a common soil-borne root parasite is *T. basicola* (Berk. & Br.) Ferr.
   (= *Milowia nivea* Massee). *Milowia* Massee (1884) would be an older name for the genus (Fig. 104b)
   References: Bateman, 1963; Barron, 1968; Kendrick and Carmichael, 1973.

4. **Sporoschisma** Berk. & Br. - Gard. Chron. p. 540 (1847)
   S. mirabile Berk. & Br.
   3 or 4 species
   References: Hughes, 1949.

5. **Acremonium** Link ex Fr. - Syst. mycol. *1*: XLIV (1821)
   A. alternatum Link ex S. F. Gray (Fig. 105a)
   = *Cephalosporium* auct. non Corda (Fig. 105b)
   ascigerous states: *Emericellopsis, Nectria*
   about 70 species, mostly described as *Cephalosporium*. A common species is the human pathogen *A. kiliense* Grütz (= *Cephalosporium acremonium* auct.)
   References: Gams, 1968, 1971.

6. **Gliomastix** Guéguen - Bull. Soc. mycol. Fr. *21*: 240 (1905)
   G. chartarum Guéguen = *G. convoluta* (Harz) Mason = *G. murorum* (Corda) Hughes (Fig. 105d)
   = *Torulina* Sacc. & D. Sacc. - Syll. Fung. *18*: 566 (1906)
   T. serotinae (Oudem.) Sacc. & D. Sacc. = *G. murorum*
   = *Basitorula* Arn. - Bull. Soc. mycol. Fr. *69*: 276 (1953)
   B. cingulata Arn. = *Gliomastix luzulae* (Fuck.) Mason & Hughes
   = *Nematomyces* Faurel & Schotter - Rev. Mycol. *30*: 330 (1965)
   N. coprophila Faurel & Schotter
   10 species
   The genus is closely related to *Acremonium*
   References: Dickinson, 1968; Gams, 1971 (as *Acremonium*).

7. **Monocillium** Saksena - Indian Phytopath. *8*: 8 (1955)
   M. indicum Saksena (Fig. 105c)
   ascigerous state: *Niesslia* Auersw.

12 species
References: Gams, 1971.

3. **Torulomyces** Delitsch - in Lembke und Delitsch: Systematik der Schimmelpilze (1943)
   T. lagena Delitsch
   1 species
   References: Barron, 1968.

9. **Phialophora** Medlar - Mycologia 7: 200 (1915)
   P. verrucosa Medlar (Fig. 105e)
   = *Cadophora* Lagerb. & Melin - Svensk Skogsvårdsför. Tidskr. *25*: 263 (1927)
   C. fastigiata Lagerb. & Melin = *Ph. fastigiata* (Lagerb. & Melin) Conant
   = *Lecythophora* Nannf. - l.c. *32*: 435 (1934)
   L. lignicola Nannf. = *Ph. lignicola* (Nannf.) Goidanich
   = *Margarinomyces* Laxa - Zentbl. Bakt. Parasitkde, Abt. 2, *81*: 392 (1930)
   M. bubakii Laxa = *Ph. bubakii* (Laxa) Schol-Schwarz
   ascigerous states: *Coniochaeta, Gaeumannomyces, Pyrenopeziza, Mollisia, Coryne*
   and other ascomycetes
   about 15 species
   References: van Beyma, 1943; Moreau, 1963; Wang, 1965; Schol-Schwarz, 1970.

10. **Rhinocladiella** Nannf. in Melin & Nannf. - Svensk Skogsvårdsför. Tidskr. *32*: 461 (1934)
   R. atrovirens Nannf. = *R. mansonii* (Castellani) Schol-Schwarz (Fig. 105f)
   = *Fonsecaea* Negroni - Rev. Inst. bact. Dep. nac. Higiene B. Aires 7: 423 (1936)
   F. pedrosoi (Brumpt) Negroni = *R. pedrosoi* (Brumpt) Schol-Schwarz
   other synonyms vide Schol-Schwarz, 1968.
   ascigerous state: *Dictyotrichiella* (known for 1 species)
   5 species, saprophytic or parasitic on man
   References: Schol-Schwarz, 1968.

11. **Exophiala** Carmichael - Sabouraudia *5*: 122 (1966)
   E. salmonis Carmichael
   second species: *E. werneckii* (Horta) v. Arx = *Pullularia werneckii* (Horta) de
   Vries = *Cladosporium werneckii* Horta
   third species: *Exophiala brunnea* Papendorf (1969).

12. **Myrioconium** Syd. - Ann. mycol. *10*: 449 (1912)
   M. scirpi Syd. = *M. scirpicola* (Ferd. & Winge) Ferd. & Winge
   = *Cristulariella* Höhnel - Sber. K. Akad. Wiss. Wien, math.-nat. Kl. *125*: 124 (1916)
   C. depraedans (Cooke) Höhnel = *M. depraedans* (Cooke) v. Arx
   = *Botryophialophora* Linder - Farlowia *1*: 403 (1944)
   B. marina Linder
   about 10 species, spermatial states of *Sclerotinia, Ciboria* or *Rutstroemia*, often
   producing sclerotia and sometimes a *Botrytis* macroconidial state
   References: Buchwald, 1949; Bowen, 1930 (for *Cristulariella*); Berthet, 1964.

13. **Stachybotrys** Corda - Icon. Fung. *1*: 21 (1837)
   S. atra Corda = *S. chartarum* (Ehrenb. ex Link) Hughes (Fig. 106a)
   = *Fuckelina* Sacc. - Nuovo G. bot. ital. 7: 326 (1875)
   F. socia Sacc. = *S. socia* (Sacc.) Sacc.

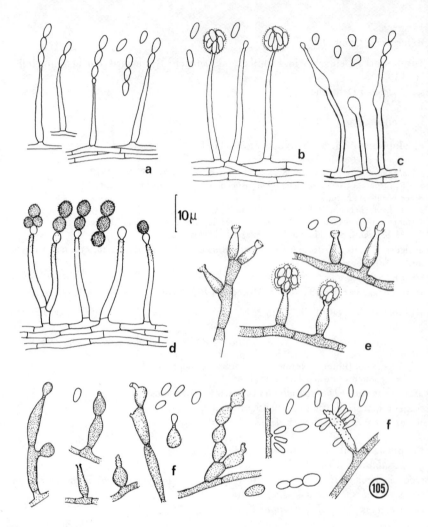

Fig. 105. a, *Acremonium alternatum*; b, *Acremonium kiliense*; c, *Monocillium indicum*; d, *Gliomastix murorum*; e, *Phialophora verrucosa*; f, *Rhinocladiella mansonii* (orig., a, b from W. Gams).

= *Hyalostachybotrys* Srinivasan - J. Indian bot. Soc. *37*: 334 (1958)
    H. bisbyi Srinivasan = *S. bisbyi* (Srinivasan) Barron
= *Synsporium* Preuss in Klotzsch Herb. viv. mycol. no. 1285 (1849)
    S. biguttatum Preuss = *S. chartarum*
ascigerous state: *Melanopsamma* (vide Booth, 1957)
about 15 species
References: Verona and Mazzucchetti, 1968; Barron, 1962; Smith, 1962; Zuck, 1946.

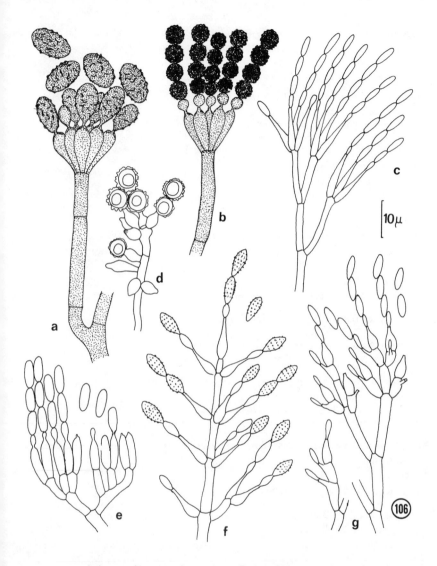

Fig. 106. a, *Stachybotrys chartarum*; b, *Memnoniella (Stachybotrys) echinata*; c, *Paecilomyces variotii*; d, *Eladia saccula*; e, *Metarrhizium anisopliae*; f, *Acrophialophora nainiana*; g, *Sesquicillium buxi* (orig., partly from W. Gams).

4. **Memnoniella** Höhnel - Zentbl. Bakt. Parasitkde, Abt. 2, *60*: 16 (1923)
   M. aterrima Höhnel = *M. echinata* (Riv.) Galloway (Fig. 106b)
   References: Verona and Mazzucchetti, 1968; Subramanian, 1971.

5. **Aspergillus** Mich. ex Fr. - Syst. mycol. *3*: 385 (1832)
   A. flavus Link ex Fr.
   = *Sterigmatocystis* Cramer - Vierteljahresschr. naturf. Ges. Zürich *4*: 325 (1859)
   S. antacustica Cramer = *A. niger* van Tiegh.

195

ascigerous states: *Eurotium, Emericella, Neosartorya, Warcupiella, Chaeto-sartorya, Hemicarpenteles, Syncleistostroma, Dichlaena*
about 160 species (Fig. 41, 42)
References: Raper and Fennell, 1965.

16. **Penicillium** Link ex Fr. - Syst. mycol. *3*: 406 (1832)
    P. expansum Link ex Fr.
    = *Citromyces* Wehmer - Ber. dt. bot. Ges. *11*: 333 (1893)
        C. glaber Wehmer = *P. glabrum* (Wehmer) Westling = *P. frequentans* Westling
    ascigerous states: *Eupenicillium* (= *Carpenteles*), *Talaromyces, Hamigera, Penicil-liopsis, Trichocoma* ≡ *Trichoskytale*
    about 220 species (Fig. 43, 44)
    References: Raper and Thom, 1949; Kulik, 1968; Stolk, 1969; Domsch and Gams. 1970.

17. **Eladia** Smith - Trans. Br. mycol. Soc. *44*: 42 (1961)
    E. saccula (Dale) G. Smith (= *Scopulariopsis verticillioides* Kamyschko) (Fig. 106d)
    1 species
    References: Barron, 1968; Domsch and Gams, 1970.

18. **Paecilomyces** Bain. - Bull. Soc. mycol. Fr. *23*: 26 (1907)
    P. variotii Bain. (Fig. 106c)
    = *Corollium* Sopp - Vidensk. Skr. Kristiania *11*: 98 (1912)
        C. dermatophagum Sopp = *P. variotii*
    = *Isaria* Pers. ex Fr. p.p.
        I. farinosa (Dicks.) Fr. = *P. farinosus* (Fr.) Brown & G. Smith
    = *Spicaria* auct. non Harting
    ascigerous states: *Byssochlamys, Thermoascus*
    References: Brown and Smith, 1957; Onions and Barron, 1967.

19. **Acrophialophora** Edward - Mycologia *51*: 784 (1959)
    A. nainiana Edward (Fig. 106f)
    3 species.
    References: Samson and Mahmood, 1970.

20. **Nomuraea** Maubl. - Bull. Soc. mycol. Fr. *19*: 295 (1903)
    N. prasina Maubl. = *N. rileyii* (Farlow) Samson
    2 species, entomogenous.

21. **Mariannaea** Arnaud ex Samson - Stud. Mycol. (in press)
    M. elegans (Corda) Samson (= *Paecilomyces elegans* (Corda) Mason & Hughes)
    4 or 5 species.

22. **Phialomyces** Misra & Talbot - Can. J. Bot. *42*: 1287 (1964)
    P. macrosporus Misra & Talbot
    1 species.

23. **Sesquicillium** W. Gams - Acta bot. neerl. *17*: 455 (1968)
    S. buxi (Link ex Fr.) W. Gams (Fig. 106g)
    3 species.

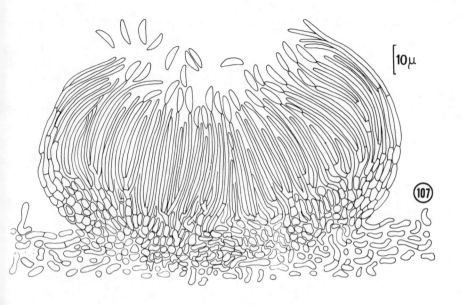

Fig. 107. *Hainesia lythri*, sporodochium (orig.).

4. **Tolypocladium** W. Gams - Persoonia *6*: 185 (1971)
    T. inflatum W. Gams
3 species
Tolypocladium is related to *Harposporium* Lohde with nemadode-trapping species, distinguishable by its narrow, curved conidia.

5. **Phialotubus** Roy & Leelavathy - Trans. Br. mycol. Soc. *49*: 495 (1966)
    P. microsporus Roy & Leelavathy
1 species.

6. **Metarrhizium** Sorokin - Les maladies des plantes, etc. *2*: 268 (1883)
    M. anisopliae (Metschn.) Sorokin (Fig. 106e)
    2 species, parasitic on insects
References: Latch, 1965; Veen, 1968.

7. **Akanthomyces** Lebert - Z. wiss. Zool. *9*: 447 (1858)
    A. aculeatus Lebert
= *Insecticola* Mains - Mycologia *42*: 577 (1950)
    I. clavata Mains
5 or 6 species, all entomogenous.

8. **Volutina** Penz. & Sacc. - Icon. Fung. javan. p. 113 (1904)
    V. concentrica Penz. & Sacc.
1 species, common on plant debris in warmer areas
References: Subramanian, 1954.

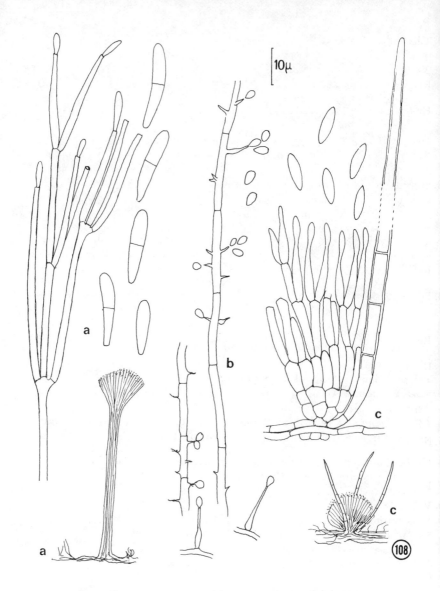

Fig. 108. a, *Didymostilbe* spec.; synnema, conidiophores with phialides and conidia; b, *Aphanocladium album*; c, *Volutella buxi*, sporodochium and part of a sporodochium with phialides and spores (orig., a from W. Gams).

9. **Aphanocladium** W. Gams - Cephalosporium-art. Schimmelpilze p. 196 (1971)
    A. album (Preuss) W. Gams (= *Acremonium album* Preuss) (Fig. 108b)
   3 species.

0. **Tilachlidium** Preuss - Linnaea *24*: 126 (1851)
   T. pinnatum Preuss = *T. brachiatum* (Batsch ex Fr.) Petch
   2 species
   References: Petch, 1937, 1945; Morris, 1963; Gams, 1971.

1. **Hirsutella** Pat. - Rev. mycol. *14*: 69 (1892)
   H. entomophila Pat.
   several species described, only a few known in pure culture
   References: Morris, 1963.

2. **Stilbella** Lindau in Engl. & Prantl - Nat. Pflanzenfam. 1, *1*: 489 (1890) (nom.
   cons.)
   S. erythrocephala (Ditm.) Lindau
   = *Stilbum* auct. non Tode ex Fr. sensu Juel
   ascigerous state: *Nectria*
   about 10 species
   References: Morris, 1963.

3. **Synnematium** Speare - Mycologia *12*: 74 (1920)
   S. jonesii Speare
   2 species
   References: Morris, 1963.

4. **Didymostilbe** P. Henn. - Hedwigia *41*: 148 (1902)
   D. coffeae P. Henn.
   = *Didymostilbe* Bres. & Sacc. - Atti Congr. bot. Palermo p. 59 (1903)
   D. eichleriana Bres. & Sacc.
   2 or 3 species (Fig. 108a).

5. **Neottiosporella** Höhnel ex Graniti - Nuovo G. bot. ital. *58*: 148 (1951)
   N. triseti Graniti
   second species: *N. radicata* Morris
   References: Morris, 1956.

6. **Volutella** Tode ex Fr. - Syst. mycol. *3*: 466 (1832)
   V. ciliata (Alb. & Schw.) Fr.
   = *Chaetodochium* Höhnel - Mitt. bot. Inst. techn. Hochsch. Wien *9*: 44 (1932)
   Ch. buxi (DC. ex Fr.) Höhnel = *V. buxi* (DC. ex Fr.) Berk. & Br. (Fig. 108c)
   ascigerous state: *Pseudonectria*
   References: Bezerra, 1963; Chilton, 1954.

7. **Nalanthamala** Subramanian - J. Indian Bot. Soc. *35*: 476 (1956)
   N. madreeya Subramanian
   2 species.

8. **Myrothecium** Tode ex Fr. - Syst. mycol. *3*: 216 (1829)
   M. roridum Tode ex Fr.
   8 species (Fig. 109a)
   References: Preston, 1943, 1948, 1961; Nicot and Olivry, 1961; Tulloch, 1972.

9. **Septomyrothecium** Matsushima - Microfungi Solomon Isl. p. 54 (1971)
   S. uniseptatum Matsushima
   1 species.

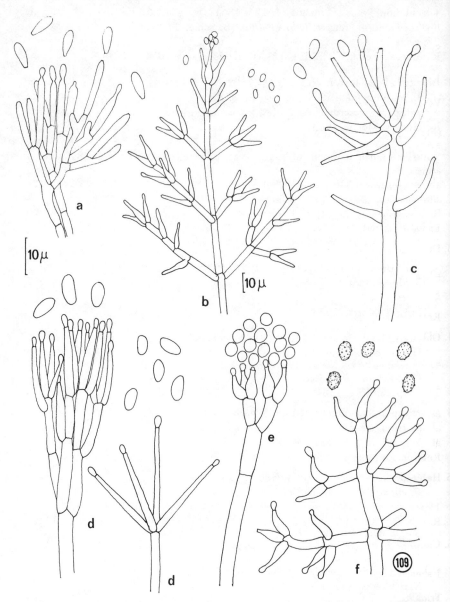

Fig. 109. a, *Myrothecium verrucaria* Ditmar ex Fr.; b, *Verticillium tenerum*; c, *Harziella capitata*; d, *Gliocladium roseum*; e, *Gliocladium virens*; f, *Trichoderma viride* (orig., partly from W. Gams).

40. **Tubercularia** Tode ex Fr. - Syst. mycol. *3*: 463 (1832)
       T. vulgaris Tode ex Fr.
   conidial state of *Nectria cinnabarina* (Tode ex Fr.) Fr.

2 or 3 species: A large number of species are described, but most of them are identical with *T. vulgaris*.

1. **Stachylidium** Link ex S.F. Gray - Nat. Arr. Br. Pl. p. 553 (1821)
   S. bicolor Link ex S.F. Gray
   1 species.

2. **Verticillium** Nees ex Link - Linn. Spec. Plant. *1*: 75 (1824)
   V. tenerum (Nees ex Pers.) Link = *V. lateritium* (Ehrenb. ex Link) Rabenh. (Fig. 109b)
   = *Acrostalagmus* Corda - Icon. Fung. *2*: 15 (1838)
      A. cinnabarinus Corda = *V. tenerum*
   ascigerous state: *Nectria*
   about 30 species, some parasitic on plants. In this respect important species are *V. albo-atrum* Reinke & Berth., *V. dahliae* Kleb. and *V. nigrescens* Pethybr.
   References: Hughes, 1951; Isaac, 1949, 1967; Isaac and Davies, 1955; van den Ende, 1958: Gams, 1971.

3. **Diheterospora** Kamyschko - Bot. Mater. *15*: 138 (1962)
   D. heterospora Kamyschko = *D. chlamydosporia* (Goddard) Barron & Onions
   = *Pochonia* Batista & Fonseca - Publ. Inst. Micol. Recife No. 462 (1965)
      P. humicola Batista & Fonseca = *D. chlamydosporia*
   2 species. The genus is hardly distinguishable from *Verticillium*.
   References: Barron and Onions, 1966; Gams, 1971 (as *Verticillium*).

4. **Gliocladium** Corda - Icon. Fung. *4*: 30 (1840)
   G. penicillioides Corda
   = *Clonostachys* Corda - Prachtflora p. 15 (1840)
      C. araucaria Corda
   = *Gibellulopsis* Batista & Maia - An. Soc. biol. Pernambuco *16*: 153 (1959)
      G. piscis Batista & Maia
   ascigerous state: *Nectria*
   about 20 species, common are *G. roseum* Bain. (Fig. 109d) and *G. virens* Miller & al. (Fig. 109e)
   References: Morquer et al., 1963; Raper and Thom, 1949; Isaac, 1954.

5. **Harziella** Cost. & Matr. (non O. Kuntze) - Bull. Soc. mycol. Fr. *15*: 104 (1899)
   H. capitata Cost. & Matr. (Fig. 109c)
   1 species. The genus is related to *Trichoderma*
   References: Lindau, 1907.

6. **Uncigera** Sacc. - Atti Ist. Ven. Sci. Lett. Arti 6,*3*: 741 (1885)
   U. cordae Sacc. & Berl.
   1 species.

. **Trichoderma** Pers. ex Fr. - Syst. mycol. *3*: 214 (1829)
   T. viride Pers. ex Fr. (Fig. 109f)
   = *Aleurisma* Link ex Fr. - Syst. mycol. *3*: 452 (1832)
      A. sporulosum Link ex Fr. = *T. polysporum* (Link ex Pers.) Rifai
   = *Pachybasium* Sacc. - Rev. mycol. 7: 160 (1885)
      P. hamatum (Bon.) Sacc. = *Trichoderma hamatum* (Bon.) Bain.
   = *Sporoderma* Mont. - Syll. Crypt. no. 1069 (1856)
      S. chlorogenum Mont.

= *Tolypomyria* Preuss - Linnaea *25*: 726 (1852)
   T. prasina Preuss = *T. viride*
about 20 species, ascigerous states of *Hypocrea* Fr., *Podostroma* Karst. and
related genera
References: Webster, 1964; Dingley, 1957; Doi, 1966, 1968; Rifai, 1969.

48. **Hainesia** Ellis & Sacc. - Syll. Fung. *3*: 699 (1884)
   H. rhoina (Sacc.) Ellis & Sacc. = *H. lythri* (Desm.) Höhnel (Fig. 107)
   ascigerous state: *Discohainesia oenotherae* (Cooke & Ellis) Nannf.
   References: von Arx, 1957, 1970.

49. **Gonytrichum** Nees ex Fr. - Syst. mycol. *3*: 348 (1832)
   G. caesium Nees ex Fr.
   = *Mesobotrys* Sacc. - Michelia *2*: 27 (1880)
   M. macroclada (Sacc.) Sacc. = *G. macrocladum* (Sacc.) Hughes (Fig. 111a)
   ascigerous state: *Melanopsammella* Höhnel (= *Trichosphaerella* Bomm. & al.)
   3 species
   References: Hughes, 1951; Swart, 1959; Barron and Bhatt, 1967.

50. **Cladorrhinum** Sacc. & March. - Bull. Soc. bot. Belg. *24*: 64 (1885)
   C. foecundissimum Sacc. & March. (Fig. 112d)
   = *Bahupaathra* Subram. & Lodha - Antonie van Leeuwenhoek *30*: 317 (1964)
   B. samala Subram. & Lodha
   ascigerous state: *Apiosordaria*
   3 or 4 species (some more are not yet described)
   References: von Arx and Gams, 1966; Gams and Domsch, 1969.

51. **Chaetopsina** Rambelli - Atti Accad. Sci. Bologna 11, *3*: 191 (1956)
   C. fulva Rambelli
   1 species.

52. **Chaetopsis** Grev. ex Corda - Icon. Fung. *1*: 11 (1837)
   C. wauchii Grev. ex Corda = *C. grisea* (Ehrenb.) Sacc.
   = *Chaetopsella* Höhnel - Mitt. bot. Lab. techn. Hochsch. Wien 7: 43 (1930)
   C. grisea (Ehrenb.) Höhnel
   1 species
   References: Hughes, 1951.

53. **Zygosporium** Mont. - Ann. Sci. nat., Bot., Sér. 2, *17*: 121 (1842)
   Z. oscheoides Mont.
   synonyms vide Mason, 1933
   References: Hughes, 1951; Wang and Baker, 1967.

54. **Thysanophora** Kendrick - Can. J. Bot. *39*: 817 (1961)
   T. penicillioides (Roum.) Kendrick
   second species: *T. longispora* Kendrick.

55. **Custingophora** Stolk & al. - Persoonia *5*: 195 (1968)
   C. olivacea Stolk & al. (Fig. 112b)
   1 species
   References: Stolk and Hennebert, 1968.

56. **Phialocephala** Kendrick - Can. J. Bot. *39*: 1079 (1961)
   P. dimorphospora Kendrick (Fig. 112c)

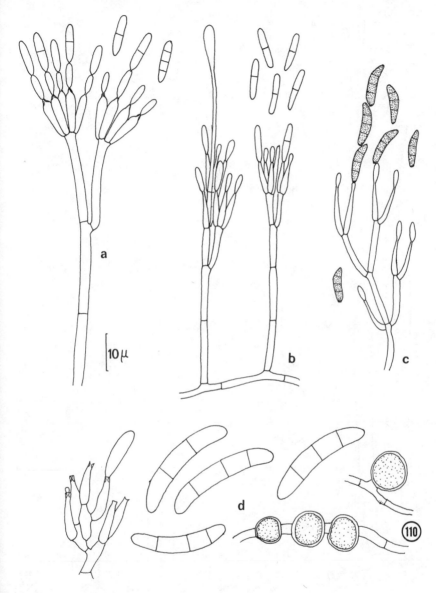

Fig. 110. a, *Penicillifer pulcher*; b, *Cylindrocladium parvum* Anderson; c, *Fusariella* spec.; d, *Cylindrocarpon olidum* Wollenw. (orig., partly from W. Gams).

3 or 4 species

References: Kendrick, 1963, 1964; Meyer, 1959; Jong & Davis, 1972; Aebi, 1972 (sub *Cystodendron* Bubak, which may be distinguished by shorter conidiophores).

7. Gliocephalis Matr. - Bull. Soc. mycol. Fr. *15*: 254 (1899)

      G. hyalina Matruchot

1 species

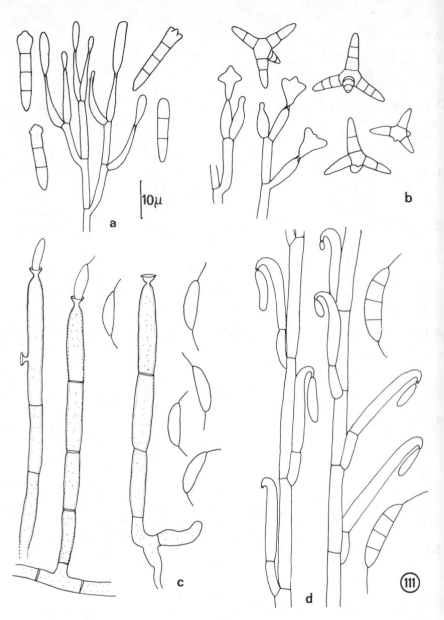

Fig. 111. a, *Heliscus lugdunensis*; b, *Lemonniera terrestris* Tubaki; c, *Codinaea simplex* Hughes & Kendrick; d, *Menispora glauca* (a, b from Tubaki, 1958, c orig., d from Hughes and Kendrick, 1963).

References: Barron, 1968.

58. **Goidanichiella** Arnaud ex Barron - Gen. Hyphom. Soil, p. 180 (1968)
     G. scopula (Preuss) Arnaud
    1 species.

9. Gibellula Cavara - Atti Ist. Bot. Pavia *2*: 347 (1894)
   G. pulchra (Sacc.) Cavara
   ascigerous state: *Torrubiella.*
   2 or 3 species, entomogenous.

0. Gliocephalotrichum J. J. Ellis & Hesseltine - Bull. Torrey bot. Club *89*: 21 (1962)
   G. bulbilium J. J. Ellis & Hesseltine
   4 species
   References: Wiley and Simmons, 1971.

1. Penicillifer van Emden - Acta bot. neerl. *17*: 54 (1968)
   P. pulcher van Emden (Fig. 110a)
   1 species.

2. Cylindrocladium Morgan - Bot. Gaz. *17*: 191 (1892)
   C. scoparium Morgan
   = *Candelospora* Hawley - Proc. R. Ir. Acad. *31* (13): 11 (1912)
      C. ilicicola Hawley = *Cylindrocladium ilicicola* (Hawley) Boedijn & Reitsma
   ascigerous state: *Calonectria*
   about 10 species (Fig. 110b)
   References: Boedijn and Reitsma, 1950; Booth and Murray, 1960; Gerlach, 1968.

3. Fusariella Sacc. - Misc. Mycol. *1*: 29 (1884)
   F. atrovirens Sacc.
   = *Kurssanovia* Pidopl. - Mykrobiol. Zh. Kiew *9*(2-3): 57 (1947)
      K. triseptata Pidopl.
   about 5 species (Fig. 110c)
   References: Hughes, 1949.

4. Fusarium Link ex Fr. - Syst. mycol. *3*: 469 (1832)
   F. roseum Link ex Fr.
   = *Pionnotes* Fr. - Summa Veg. Scand. p. 481 (1849)
   ascigerous state: *Nectria, Calonectria, Gibberella, Nectriopsis, Griphosphaeria,*
   Plectosphaerella
   about 40 species (Fig. 64)
   References: Wollenweber and Reinking, 1935; Gordon, 1952; Booth, 1959, 1960,
   1971; Seemüller, 1968; Gerlach, 1970; Gerlach and Ershad, 1970.

5. Cylindrocarpon Wollenw. - Phytopathology *3*: 225 (1913) (nom. cons.)
   C. cylindroides Wollenw.
   = *Allantospora* Wakker - Meded. Proefst. Oost Java 2, *28*: 8 (1896)
      A. radicicola Wakker = *C. didymum* (Hartig) Wollenw.
   = *Coleomyces* Moreau - Bull. Soc. mycol. Fr. *53*: 33 (1937)
      C. rufus Moreau = *C. destructans* (Zins.) Scholten
   = *Fusidium* Link ex Fr. - Syst. mycol. *3*: 480 (1832)
      F. candidum Link ex Fr. = *C. candidum* (Link ex Fr.) Wollenw.
   = *Moeszia* Bubák - Bot. Koezlem. *13*: 94 (1914)
      M. cylindroides Bubák = *C. magnusianum* Wollenw.
   ascigerous state: *Nectria*
   about 30 species (Fig. 110d)
   References: Booth, 1959, 1966; Gerlach, 1956, 1959.

66. **Heliscus** Sacc. - Michelia 2: 1 (1880)
 H. lugdunensis Sacc. & Therry (Fig. 111a)
 ascigerous state: *Nectria*
 References: Nilsson. 1964; Ingold and Cox, 1957; Webster, 1959.

67. **Clavatospora** Nilsson - Symb. bot. upsal. *18*, 2: 88 (1964)
 C. longibrachiata (Ingold) Nilsson
 3 species.

68. **Lemonniera** de Wildeman - Ann. Soc. belge Micr. *18*: 143 (1894)
 L. aquatica de Wildeman
 4 or 5 species
 References: Nilsson, 1964; Tubaki, 1958 (Fig. 111b).

69. **Alatospora** Ingold - Trans. Br. mycol. Soc. *25*: 381 (1942)
 A. acuminata Ingold
 1 species.

70. **Anguillospora** Ingold - Trans. Br. mycol. Soc. *25*: 398 (1942)
 A. longissima (Sacc. & Syd.) Ingold
 4 species
 References: Nilsson, 1964.

71. **Articulospora** Ingold - Trans. Br. mycol. Soc. *25*: 372 (1942)
 A. tetracladia Ingold
 4 species
 References: Peterson, 1962.

72. **Codinaea** Maire - Publ. Inst. bot. Barcelona *3*: 15 (1937)
 C. aristata Maire
 = *Menisporella* Agnihotrudu - Proc. Indian Acad. Sci., suppl. B, *56*: 97 (1963)
  M. assamica Agnihotrudu = *Codinaea assamica* (Agnihotrudu) Hughes & Kendrick
 16 or 17 species (Fig. 111c)
 References: Hughes and Kendrick, 1963, 1968.

73. **Menispora** Pers. ex Fr. - Syst. mycol. *3*: 451 (1832)
 M. glauca Pers. ex Fr. (= *Psilonia glauca* Fr.) (Fig. 111d)
 ascigerous state: *Chaetosphaeria* Tul.
 6 or 7 species
 References: Hughes and Kendrick, 1963, 1968.

74. **Menisporopsis** Hughes - Mycol. Pap. *48*: 59 (1952)
 M. theobromae Hughes
 1 species
 References: Meyer, 1959.

75. **Chloridium** Link ex Sacc. - Syll. Fung. *4*: 320 (1886)
 C. viride Link ex Sacc.
 = *Bisporomyces* van Beyma - Antonie van Leeuwenhoek *6*: 277 (1940)
  B. chlamydosporis van Beyma = *C. chlamydosporis* (van Beyma) Hughes (Fig. 113a)
 = *Cirrhomyces* Höhnel - Ann. mycol. *1*: 529 (1903)
  C. caudiger Höhnel = *Chloridium caudigerum* (Höhnel) Hughes
 = *Psilobotrys* Sacc. - Michelia *1*: 538 (1879)
  P. minuta Sacc. = *Chloridium minutum* (Sacc.) Sacc.

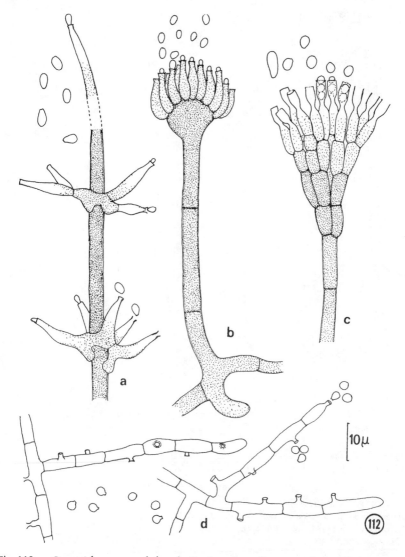

Fig. 112. a, *Gonytrichum macrocladum*; b, *Custingophora olivacea*; c, *Phialocephala dimorphospora*; d, *Cladorrhinum foecundissimum* (a, c, d orig. from W. Gams, b orig. from A. C. Stolk).

ascigerous state: *Chaetosphaeria* Tul.
4 or 5 species, some of them described as *Catenularia*
References: Hughes, 1958; Barron, 1968.

6. **Catenularia** Grove ex Sacc. - Syll. Fung. *4*: 303 (1886)
    C. simplex Grove = *C. cuneiformis* (Richon) Mason (Fig. 113b)
= *Psiloniella* Costantin - Les Mucédinées simples p. 190 (1888)
ascigerous state: *Chaetosphaeria* Tul.

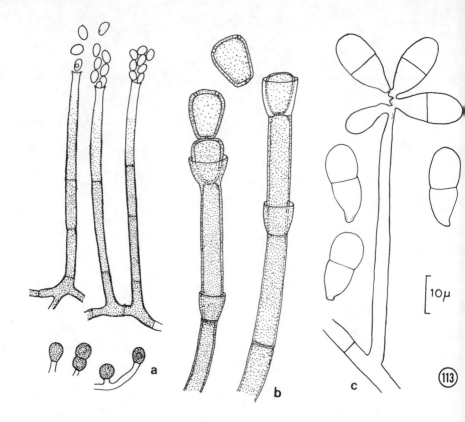

Fig. 113. a, *Chloridium chlamydosporis*; b, *Catenularia cuneiformis*; c, *Trichothecium roseum* (b from Booth, 1958, a, c orig.).

4 species
References: Hughes, 1965; Booth, 1957, 1958.

77. **Scopulariopsis** Bain. - Bull. Soc. mycol. Fr. *23*: 98 (1907)
    S. brevicaulis (Sacc.) Bain. (Fig. 114a)
  = *Acaulium* Sopp - Skr. Vidensk. Selsk. Crist. *11*: 42 (1912)
    A. nigrum Sopp = *S. asperula* (Sacc.) Hughes
  = *Masonia* G. Smith - Trans. Br. mycol. Soc. *35*: 149 (1952) non Hansford (1944)
  ≡ *Masoniella* G. Smith - l.c. *35*: 237 (1952)
    M. grisea G. Smith = *S. brumptii* Salvanet-Duval
  = *Phaeoscopulariopsis* Ota - Jap. J. Derm. Urol. *28*: 405 (1928)
ascigerous state: *Microascus*
about 30 species
References: Morton and Smith, 1963.

78. **Cephalotrichum** Link ex Fr. - Syst. mycol. *3*: 280 (1829)
    C. stemonitis (Pers.) Link ex Fr.
  = *Doratomyces* Corda in Sturm Deutschl. Fl., Pilze 3, *2*: 65 (1829)
    D. neesii Corda = *C. stemonitis*
  = *Stysanus* Corda - Icon. Fung. *1*: 22 (1837)
    S. stemonitis (Pers.) Corda = *C. stemonitis*

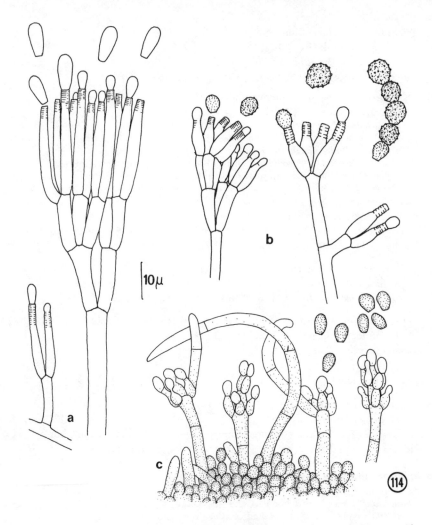

Fig. 114. a, *Leptographium lundbergii*; b, *Scopulariopsis brevicaulis*; c, *Trichurus spiralis* Hasselbring (a, b orig., c from Domsch and Gams, 1970).

References: Hughes, 1958; Morton and Smith, 1963; Swart, 1964. Morton and Smith and others are using the genus name *Doratomyces*, but there is no reason to reject the older name *Cephalotrichum!*

9. **Trichurus** Clem. & Shear - Bot. Surv. Nebr. *4*: 7 (1896)
    T. cylindricus Clem. & Shear
4 or 5 species (Fig. 114c)
References: Swart, 1964; Bainier, 1907; Lodha, 1963.
The genus *Trichurus* is hardly distinguishable from *Cephalotrichum*.

80. **Trichothecium** Link - Linn. Spec. Pl. *1*: 28 (1824)
    T. roseum Link (Fig. 113a)
    ascigerous state: *Hypomyces* Tul.(?)
    References: Ingold, 1956; Tubaki, 1958; Meyer, 1958; Rifai, 1965.

81. **Cladobotryum** Nees ex Steudel - Nomencl. bot. p. 118 (1824)
    C. varium Nees ex Steudel
    = *Didymocladium* Sacc. - Syll. Fung. *4*: 186 (1886)
       D. ternatum (Bon.) Sacc. = *C. varium* Nees
    = *Diplocladium* Bon. - Handb. allg. Mykol. p. 98 (1851)
       D. major Bon. ? = *C. mycophilum* (Oudem.) W. Gams & Hoozemans
    = *Dactylium* Nees sensu Sacc. - Michelia *2*: 20 (1880)
       D. dendroides (Bull.) Fr. = *C. dendroides* (Bull. ex Mérat) W. Gams & Hoozemans (Fig. 67)
    ascigerous state: *Hypomyces*
    6 or 7 species, on decaying mushrooms
    References: Lentz, 1967 (as *Dactylium*); Tubaki, 1955 (as *Monosporium*);
    Hughes, 1958; Gams and Hoozemans, 1970.

82. **Sibirina** Arnold - Nova Hedwigia *19*: 299 (1970)
    S. fungicola Arnold
    Second species: *S. orthospora* W. Gams - Persoonia 7: 163 (1973)

83. **Leptographium** Lagerb. & Melin - Svensk Skogsvårdsför. Tidskr. *25*: 248 (1927)
    L. lundbergii Lagerb. & Melin (Fig. 114a)
    ascigerous state: *Petriella* (?)
    1 or 2 species
    References: Hughes, 1953; Meyer, 1959; Shaw & Hubert, 1952.

84. **Graphium** Corda - Icon. Fung. *1*: 18 (1837)
    G. penicillioides Corda
    In the type species, the conidiogenous cells may show annellations, but may also
    elongate sympodially; the base of the conidia is truncate.

85. **Spilocaea** Fr. - Syst. mycol. *3*: 504 (1832)
    S. pomi Fr. (Fig. 74)
    = *Basiascum* Cavara - Atti Ist. bot. Pavia 2, *1*: 433 (1888)
       B. eriobotryae Cavara = *S. eriobotryae* (Cavara) Hughes
    = *Cycloconium* Cast. - Catal. pl. Marseilles p. 220 (1845)
       C. oleagineum Cast. = *S. oleaginea* (Cast.) Hughes
    = *Napicladium* Thüm. - Mycotheca univ. no. 91 (1875)
       N. soraueri Thüm. = *S. pomi*
    ascigerous state: *Venturia*
    4 or 5 species, parasitic on higher plants
    References: Hughes, 1953; von Arx, 1957.

86. **Pollaccia** Bald. & Cif. - Atti Ist. bot. Univ. Pavia 4, *10*: 55 (1938)
    P. radiosa (Lib.) Bald. & Cif.
    3 species, parasitic on *Populus* and *Salix*
    References: Servazzi, 1939; von Arx, 1957.

87. **Polypaecilum** G. Smith - Trans. Br. mycol. Soc. *44*: 437 (1961)
    P. insolitum G. Smith

ascigerous state: *Dichotomomyces* (Fig. 45)
2 species.

8. **Basipetospora** Cole & Kendrick - Can. J. Bot. *46*: 991 (1968)
   B. rubra Cole & Kendrick, a name introduced for the conidial state of *Monascus ruber*.

Moniliales Key IV (genera with blastoconidia)

1. Conidiophores arise from ampulliform cells, are narrow, cylindrical, septate, conidia lateral or terminal, dark, lenticular, cornute or flattened, 1-celled or cruciately septate, not catenulate 2

1. Not combining above characters 5

2. Conidia 1-celled 3

2. Conidia 4- or multicelled 4

3. Conidiophores with many darkened septa .... *Arthrinium* (117)

3. Conidiophores without darkened septa .... *Papularia* (118)

4. Conidia lateral and terminal, without projections... *Dictyoarthrinium* (116)

4. Conidia only terminal, lobed or with projections.......... cf. *Spegazzinia*

5. Conidia form acropetal chains 6

5. Conidia not catenulate 29

6. Conidia spherical, dark, usually verrucose, chains short, often branched.....
   ....*Periconia* (19)

6. Not combining above characters 7

7. Conidia dark, multiseptate, often readily breaking up into its separate cells, conidiogenous cells inconspicuous, usually darkened, verrucose. *Torula* (102)

7. Not combining above characters 8

8. Conidial chains arise from intercalary and terminal swellings 9

8. Conidial chains not arising from intercalary swellings 10

9. Conidia and conidiophores pigmented .... *Gonatobotryum* (30)

9. Conidia and conidiophores hyaline................ *Nematogonium* (29)

10. Hyphae and conidiophores wide, mostly brightly coloured, conidia 1-celled, spherical, barrel-shaped or ellipsoidal 11

10. Hyphae and conidiophores usually 2-4 μm wide or dark when wider 13

11. Conidiophores long, branched or swollen apically, yellow, conidia barrel-shaped, with separating cells .... *Amblyosporium* (33)

11. Not combining above characters 12

12. Conidiophores often in pustules or clustered, chains long, conidia with or without disjunctors, often barrel-shaped................... *Monilia* (1)

12. Conidiophores usually thick-walled, conidia ellipsoidal, in short chains .....
    .... *Haplotrichum* (2)

13. Synnemata present (especially in fresh cultures) 14

13. Synnemata absent 15

14. Conidia roundish or ellipsoidal, without scars, bright, in short, often lateral chains .... *Mycosylva* (17)

212

14. Conidia elongated, with scars, dark, in usually terminal chains ....
.... *Pycnostysanus* (12)

15. Conidia and conidiogenous cells usually hyaline.................... 16

15. Conidia and conidiogenous cells pigmented........................ 22

16. Conidia with usually prominent scars, dry........................ 17

16. Conidia without scars, apiculate or rounded....................... 18

17. Conidia cylindrical, usually smooth (parasitic or saprophytic on leaves) .....
.... *Ramularia* (4)

17. Conidia ellipsoidal, usually verrucose, in long chains (saprophytic) ....
.... *Hyalodendron* (6)

18. Conidiophores inconspicuous, short, conidia cylindrical, in long chains, often
forming pustulate sporodochial tufts .... *Cylindrium* (5)
(Sporodochia orange or yellow: cf. *Cylindrocolla*)

18. Conidiophores usually distinct, sporodochia absent.................. 19

19. Conidiophores upright, pigmented, conidia cylindrical, dry ....
.... *Polyscytalum* (11)

19. Conidiophores hyaline or absent, conidia usually not cylindrical........ 20

20. Colonies yeast-like, restricted in growth (parasitic in higher plants).......
.... cf. *Ustilago*

20. Colonies not yeast-like (often isolated from animals or agarics)........... 21

21. Isolated from agarics (mushrooms)................ *Myceliophthora* (15)

21. Isolated from lungs of animals .... *Emmonsia* (16)
(Isolated from other sources: cf. *Sporothrix*)

22. Conidia 1-celled or with transverse septa only, not beaked 23

22. Conidia usually with transverse and longitudinal septa or beaked 116

23. Conidiophores and conidiogenous cells inconspicuous.............. 24

23. Conidiophores usually upright and branched 26

24. Conidia 1-celled, in long chains....................... *Xylohypha* (7)

24. Conidia septate 25

25. Conidia 2-celled, with darkened septa..................... *Bispora* (8)

25. Conidia 2- or many-celled, septa not darkened .... *Septonema* (9)

26. Conidia without prominent scars, often papillate, conidiophores upright,
apically branched, colonies chestnut brown (saprophytic). . *Hormoconis* (13)

26. Conidia with prominent scars or truncate ends 27

27. Conidia smooth, 1-celled, with truncate ends .... *Alysidium* (14)

27. Conidia with prominent scars, or borne on pores 28

28. Conidia usually verrucose, small or large, 1- or many-celled . . . .
. . . . *Cladosporium* (10)

28. Conidia usually multicelled, smooth, scars with pores 120

29. Budding conidia present, colonies yeast-like. 30

29. Budding conidia usually absent, colonies not yeast-like 31

30. Colonies bright, usually restricted . . . . cf. Torulopsidales

30. Colonies spreading, often becoming dark by pigmented hyphae or cells, conidia often borne simultaneously on hyphae or on swollen branches . . . .
. . . . *Aureobasidium* (34)

31. Conidia borne simultaneously. 33

31. Conidia borne successively, usually in sympodulae, rarely singly 45

32. Conidia multicelled . . . . *Cephaliophora* (31)

32. Conidia 1-celled 33

33. Conidia small, spherical or nearly so, conidiogenous cells roundish in outline, borne on denticles on sympodially elongating conidiophores, colonies light
. . . . *Blastobotrys* (35)

33. Not combining above characters 34

34. Conidiogenous cells intercalary, swollen. . . . . *Gonatobotrys* (18)

34. Conidiogenous cells usually not intercalary 35

35. Conidiophores simple, with a spherical or clavate conidiogenous head. . . . 36

35. Conidiophores usually branched or bearing several conidiogenous cells . . . 37

36. Conidia borne on scars . . . . *Oedocephalum* (32)

36. Conidia borne on denticles or projections . . . . cf. *Spiniger*

37. Conidiophores pyramidally branched, long . . . . *Botryosporium* (28)

37. Conidiophores not pyramidally branched 38

38. Conidiophores moniliform, with lateral or catenate conidiogenous cells . . . . .
. . . . *Phymatotrichopsis* (27)

38. Conidiophores upright, often pigmented. 39

39. Conidiogenous cells elongated, usually cylindrical or tapering above. 40

39. Conidiogenous cells spherical, clavate or lobed 41

40. Conidiogenous cells sterile in the apical tapering part, conidia reticulate . . . .
. . . . *Pulchromyces* (26)

40. Conidiogenous cells apically rounded and fertile . . . . *Chromelosporium* (25)
(compare also *Mycotypha*, Mucorales)

41. Conidiophores dichotomously branched 42

41. Conidiophores especially in the upper part racemosely or irregularly branched 43

42. Sclerotia present, conidiogenous cells lobed or irregular in shape, conidia spherical .... *Amphobotrys* (22)

42. Sclerotia absent, conidiogenous cells spherical, conidia apiculate ....
.... *Dichobotrys* (24)

43. Conidiogenous cells swollen, spherical or broadly clavate, conidia smooth. . .
.... *Botrytis* (20)

43. Conidiogenous cells unswollen, often with short branches or lobes, conidia smooth or verrucose 44

44. Conidia smooth, spherical, small sclerotia present....... *Streptobotrys* (21)

44. Conidia verrucose, pyriform, brown, true sclerotia absent ....
.... *Verrucobotrys* (23)

45. Conidia curved, reniform, falcate, forked, Y-shaped, dendroid, biconical, lageniform, subulate, regularly constricted or appendaged 46

45. Conidia usually straight, not appendaged, without regular constrictions . . 67

46. Conidia biconical or lageniform-turbinate, pigmented, with a light equatorial band 47

46. Conidia not biconical or lageniform when pigmented 48

47. Conidia lageniform-turbinate, often caudate .... *Beltraniella* (81)

47. Conidia biconical .... *Beltrania* (80)
(and some more genera not distinguishable in pure culture)

48. Conidia tapering from the truncate base to an apical, linear appendage, conidiogenous cell integrated, with wide denticles .... *Subulispora* (76)

48. Conidia not subulate, not tapering.............................. 49

49. Conidia fusiform or falcate, 1-celled, with regular constrictions........ 50

49. Conidia without such constrictions when fusiform-falcate 52

50. Conidiogenous cells borne on hyphae .... *Isthmolongispora* (78)

50. Conidiogenous cells borne on erect conidiophores 51

51. Setose sporodochia present..................... *Wiesneriomyces* (79)

51. Sporodochia absent .... *Speiropsis* (77)

52. Conidia reniform, pigmented, 1-celled 53

52. Conidia not reniform when 1-celled............................ 54

53. Conidiogenous cells without denticles, conidia often with germslits or verrucose .... *Melanographium* (64)

53. Conidiogenous cells denticulate, conidia without germslits.... *Virgaria* (63)

54. Conidia 1-2-celled, hyaline, reniform, falcate or Y-shaped, in this case rarely multicelled 55

54. Conidia multicelled, hyaline or pigmented, often with appendages...... 58

55. Conidia Y-shaped, colonies dark .... *Scolecobasidium* (70)

81. Conidia with a rounded, sometimes inconspicuously apiculate base, usually small, 1-celled, hyaline. 82

81. Conidia with a pointed, conspicuously apiculate or slightly truncate base, hyaline or pigmented 85

82. Conidiogenous cells in whorls, long cylindrical or tapering, becoming zig-zag, without denticles .... *Tritirachium* (47)

82. Not combining above characters. 83

83. Conidiogenous cells with a zig-zag, denticulate rachis. .... *Beauveria* (48)

83. Not combining above characters. 84

84. Synnemata usually present, conidiogenous cells apically with a short, denticulate rachis .... *Isaria* (41)

84. Synnemata usually absent, conidiogenous cells flask-shaped, apically with a short, unilateral rachis. .... *Costantinella* (43)

85. Conidiogenous cells hyaline, conidia small, 1(-2)-celled, hyaline, synnemata absent 86

85. Conidiogenous cells usually pigmented or conidia septate, synnemata may be present 89

86. Conidiogenous cells arise from differentiated conidiophores. 87

86. Conidiogenous cells arise from undifferentiated hyphae 88

87. Conidiogenous cells umbellately arranged at the apex of the conidiophore, conidia 1-2-celled .... *Pseudobotrytis* (42)

87. Conidiogenous cells arranged in whorls .... *Calcarisporium* (49)

88. Conidiogenous cells cylindrical, apically with a small number of cylindrical, sterigmata-like denticles .... *Dexhowardia* (37)

88. Conidiogenous cells usually with numerous, short denticles, conidia often clustered, sometimes in short chains .... *Sporothrix* (36)

89. Conidia small, pointed or slightly truncate at the base, guttuliform or fusiform, 1(-2)-celled. 90

89. Conidia ellipsoidal, spherical or cylindrical, often septate 93

90. Synnemata with an apical brush present, denticles absent or inconspicuous, conidiogenous cells sympodial or percurrent. . *Graphium* (39), *Pesotum* (40)

90. Not combining above characters 91

91. Conidiogenous cells tapering towards the apex, with a long rachis .... *Acrodontium* (50)

91. Conidiogenous cells not tapering, usually cylindrical 92

92. Denticles long, often sterigmata-like, synnemata often present .... *Phaeoisaria* (51)

92. Denticles small, scar-like, phialides often also present .... cf. *Rhinocladiella*

List of genera

(General references: Barron, 1968; Ellis 1971;
Subramanian, 1971; Kendrick and Carmichael, 1973)

1. **Monilia** Pers. ex Fr. - Syst. mycol. *3*: 409 (1832)
   M. fructigena Pers. ex Fr. (l.c. p. 430 sub *Oidium*) (Fig. 115c)
   ascigerous states: *Monilinia, Neurospora*
   about 20 species.

2. **Haplotrichum** Link - Linn. Spec. Plant *1*: 52 (1824)
   H. capitatum Link
   basidial state: *Botryobasidium*

2 or 3 species
References: Linder, 1942 (as *Oidium candicans*); Holubová-Jechová, 1969 (as *Oidium*)*.

3. **Ovularia** Sacc. - Michelia 2: 17 (1880)
   O. obovata (Fuckel) Sacc. = *O. obliqua* (Cooke) Sacc. (Fig. 115a)
   ascigerous state: *Mycosphaerella*
   about 80 species, causing leaf spots. In some species the conidia are septate, e.g. in *O. decipiens* Sacc. = *O. simplex* Pass. = *Didymaria didyma* auct. (non *Ramularia didyma* Unger).

4. **Ramularia** Unger - Exanth. Pfl. p. 169 (1833) sensu Sacc. - Michelia 2: 20 (1880)
   R. didyma Unger (= *R. aequivoca* (Ces.) Sacc.) (Fig. 115b)
   = *Didymaria* Corda - Icon. Fung. 6: 8 (1854)
      D. ungeri Corda = *Ramularia didyma* Unger
   ascigerous state: *Mycosphaerella*
   about 250 species are described, mostly as the cause of leaf spots. Most likely the fungus studied by Unger was a *Ramularia* sensu Saccardo, viz. *R. aequivoca*. The fungus now known as *Didymaria didyma* (Unger) Schroet., however, is different and has to be classified as *Ovularia decipiens* Sacc. (but compare Hughes, 1949).

5. **Cylindrium** Bon. - Handb. allg. Mykol. p. 34 (1851)
   C. flavovirens Bon. = *Fusidium aeruginosum* Link. ex Fr. = *Cylindrium aeruginosum* (Link ex Fr.) Lindau
   3 or 4 species
   References: Donk, 1964.

6. **Hyalodendron** Diddens - Zentbl. Bakt. Parasitkde, Abt. II., 90: 315 (1934)
   H. lignicola Diddens
   second species: *H. album* (Dowson) Diddens.

7. **Xylohypha** (Fr.) Mason apud Deighton - Mycol. Pap. 78: 43 (1960)
   X. nigrescens (Pers. ex Fr.) Mason
   1 or 2 species.

8. **Bispora** Corda - Icon. Fung. 1: 9 (1837)
   B. monilioides Corda = *B. antennata* (Pers. ex Fr.) Mason
   3 or 4 species (Fig. 115e)
   References: Hughes, 1953.

9. **Septonema** Corda - Icon. Fung. 1: 9 (1837)
   S. secedens Corda
   6 or 7 species (Fig. 115d)
   References: Hughes, 1952, 1953.

10. **Cladosporium** Link ex Fr. - Syst. mycol. 3: 368 (1832)
    C. herbarum Link ex Fr. (Fig. 75)
    = *Biharia* Thirum. & Mishra - Sydowia 7: 79 (1953)
       B. vangueriae Thirum. & Mishra = *C. vangueriae* (Thirum. & Mishra) v. Arx comb. nov.

* The genus name *Oidium* generally is in use for the conidial states of *Erysiphe* and related genera. A more suitable name for these would be *Acrosporium* Nees ex S. F. Gray. *A. monilioides* Nees ex F. S. Gray is the conidial state of *Erysiphe graminis* DC. ex Mérat.

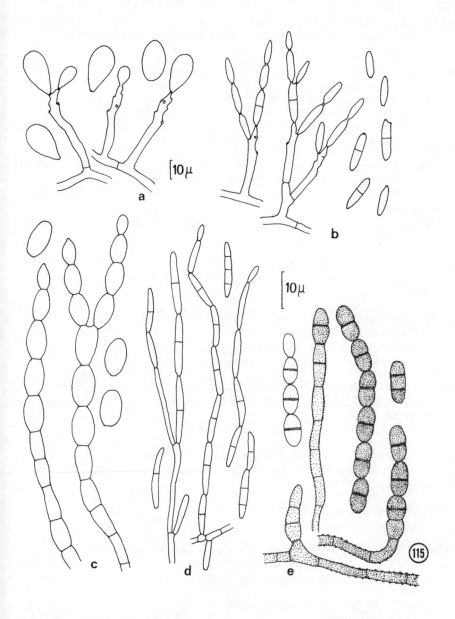

Fig. 115. a, *Ovularia obliqua*; b, *Ramularia didyma*; c, *Monilia fructigena*; d, *Septonema chaetospira* (Grove) Hughes; e, *Bispora pusilla* Sacc. (orig., d from W. Gams, e from A. C. Stolk).

= *Fulvia* Ciferri - Atti Ist. bot. Univ. Lab. critt. Pavia 5,*10*: 245 (1954)
    C. *fulvum* Cooke
= *Fusicladiopsis* Karakulin & Vasil. - Paras. nes. griby p. 193 (1937)
    F. cerasi (Rabenh.) Karakulin & Vasil. = *C. cerasi* (Rabenh.) Aderh.
≡ *Karakulinia* Golovina - Nov. Syst. Plant non vasc. p. 212 (1964)

= *Heterosporium* Cooke - Grevillea *5*: 122 (1877)
   H. ornithogali (Klotzsch) Cooke = *C. ornithogali* (Klotzsch) de Vries
= *Hormodendrum* auct. non. Bon. - Handb. allg. Mykol. p. 76 (1851)
   (*H. olivaceum* (Corda) Bon. = *Penicillium* spec.)
= *Mycovellosiella* Rangel - Arch. Jard. bot. Rio de Jan. *2*: 71 (1917)
   M. cajani (P. Henn.) Rangel
= *Phaeoramularia* Muntañola - Lilloa *30*: 182 (1960)
   P. gomphrenicola (Speg.) Muntañola
= *Stenella* Syd. - Ann. mycol. *28*: 205 (1930)
   S. araguata Syd. = *C. araguatum* (Syd.) v. Arx comb. nov.
ascigerous states: *Mycosphaerella, Venturia*
about 35 species; common saprophytic species are *C. herbarum, C. clado-sporioides* (Fres.) de Vries and *C. sphaerospermum* Penz. Some leaf spot causing species have been classified in *Heterosporium* and other genera and may be related to *Fusicladium.*
References: de Vries, 1952; Litvinov, 1967; Jacques, 1941; Ondrej, 1971; Yamamoto, 1959.

11. **Polyscytalum** Riess - Bot. Ztg. *11*: 138 (1853)
    P. fecundissimum Riess
    1 species.

12. **Pycnostysanus** Lindau - Abh. bot. Ver. Brandenb. *45*: 160 (1903)
    P. resinae (Fr.) Lindau
    second species: *P. azaleae* (Peck) Mason
    Sorocybe Fr. may be an older genus name.

13. **Hormoconis** v. Arx & de Vries - Verh. K. ned. Akad. Wet. 2, *61*: 62 (1973)
    H. resinae (Lindau) v. Arx & de Vries = *Cladosporium resinae* (Lindau) de Vries
    ascigerous state: *Amorphotheca resinae* Parbery.

14. **Alysidium** Kunze ex Steudel - Nomencl. bot. p. 54 (1824)
    A. fulvum Kunze ex Steudel = *A. dubium* (Pers. ex Fr.) M.B. Ellis
    References: Ellis, 1971.

15. **Myceliophthora** Cost. in Cost. & Matr. - Rev. gen. Bot. *6*: 289 (1894)
    M. lutea Cost., parasitic on mushrooms.

16. **Emmonsia** Cif. & Montem. - Mycopath. Mycol. appl. *10*: 303 (1959)
    E. parva (Emmons & Ashburn) Cif. & Montem.
    second species: *E. crescens* Emmons & Jellison, both parasitic in lungs of rodents.

17. **Mycosylva** Tulloch - Trans. Br. mycol. Soc. *60*: 155 (1973)
    M. clarkii Tulloch
    second species: *M. reticulata* Samson & Hintikka

18. **Gonatobotrys** Corda - Prachtflora p. 9 (1839)
    G. simplex Corda (Fig. 116b)
    2 species
    References: Drechsler, 1950; Swart, 1959.

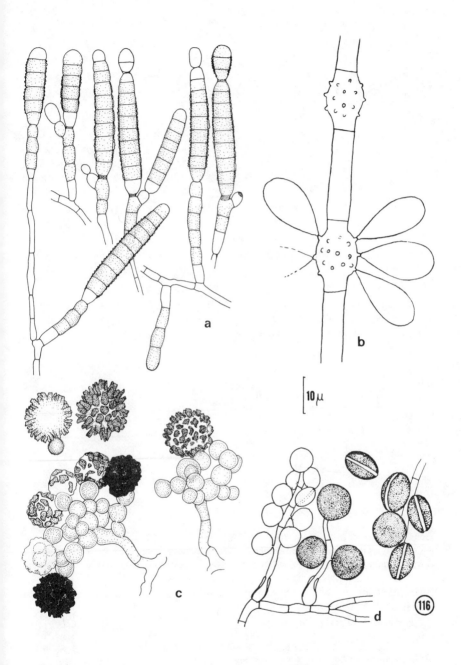

10μ

Fig. 116. a, *Dendryphion nanum*; b, *Gonatobotrys simplex* Corda; c, *Periconia macrospinosa* Lefebvre & Johnson; d, *Papularia phaeosperma* (a and c from Domsch and Gams, 1970, b and d orig.).

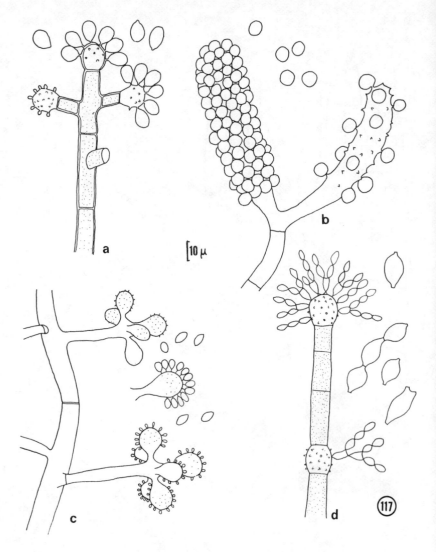

Fig. 117. a, *Botrytis cinerea*; b, *Chromelosporium* conidial state of *Peziza ostracoderma* c, *Botryosporium pulchrum*; d, *Gonatobotryum apiculatum* (Peck) Hughes (orig.).

19. **Periconia** Tode ex Schw. - Syn. Fung. Carol. Sup. p. 125 (1822) ex Fr. - Syst. mycol. *3*: 307 (1832)

    P. byssoides Pers. ex Schw. (neotype)
= *Sporodum* Corda - Icon. Fung. *1*: 18 (1829)
    S. conopleoides Corda = *P. hispidula* (Pers.) Mason & M. B. Ellis
= *Trichocephalum* Cost. - Les Mucédinées simples p. 106 (1888)
    T. curtum (Berk.) Cost. = *P. curta* (Berk.) Mason & M. B. Ellis
about 30 species (Fig. 116c)
References: Mason and Ellis, 1953; Subramanian, 1955; Rao and Rao, 1964.

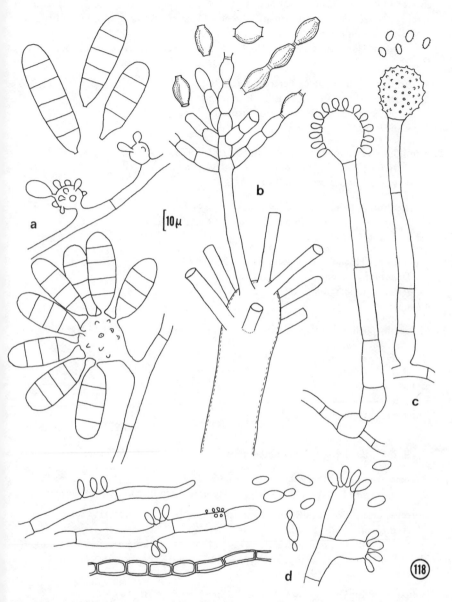

Fig. 118. a, *Cephaliophora tropica*; b, *Amblyosporium spongiosum*; c, *Oedocephalum beticola* Oud.; d, *Aureobasidium pullulans* (orig., b from W. Gams).

). Botrytis Pers. ex Fr. - Syst. mycol. *3*: 393 (1832)
      B. cinerea Pers. ex Fr. (Fig. 117a)
   ascigerous state: *Sclerotinia* (often described as *Botryotinia*)

about 30 species (200 described as herbarium specimens), mostly parasitic on higher plants. The genus urgently needs revision. A large number of species, especially the older ones, have to be excluded.
References: Hennebert, 1963, 1964; Hennebert and Groves, 1963; Menzinger, 1966.

21. **Streptobotrys** Hennebert - Persoonia 7: 191 (1973)
    S. streptothrix (Cooke & Ellis) Hennebert
    ascigerous state: *Sclerotinia (Streptotinia)*
    2 species.

22. **Amphobotrys** Hennebert - Persoonia 7: 192 (1973)
    A. ricini (Buchwald) Hennebert = *Botrytis ricini* Buchwald
    ascigerous state: *Sclerotinia (Botryotinia).*

23. **Verrucobotrys** Hennebert - Persoonia 7: 193 (1973)
    V. geranii (Seaver) Hennebert = *Botrytis geranii* Seaver
    ascigerous state: *Sclerotinia (Seaverinia).*

24. **Dichobotrys** Hennebert - Persoonia 7: 193 (1973)
    D. abundans Hennebert
    ascigerous states: *Trichophaea, Sphaerosporella*
    4 species.

25. **Chromelosporium** Corda in Sturm, Deutschl. Fl., Pilze 3,*3*: 81 (1833)
    Ch. ochraceum Corda
    = *Dissoacremoniella* Kirilenko - Nov. Syst. Pl. non vasc. 7: 235 (1970)
      D. silvatica Kirilenko
    ascigerous state: *Peziza*
    9 species
    References: Hennebert, 1973.

26. **Pulchromyces** Hennebert - Persoonia 7: 191 (1973)
    P. fimicola (Dring) Hennebert = *Phymatotrichum fimicola* Dring.

27. **Phymatotrichopsis** Hennebert - Persoonia 7: 199 (1973)
    P. omnivora (Duggar) Hennebert = *Phymatotrichum omnivorum* Duggar.

28. **Botryosporium** Corda in Sturm, Deutschl. Fl., Pilze 3, *3*: 9 (1831)
    B. diffusum (Alb. & Schw.) Corda
    other species: *B. pulchrum* Corda (Fig. 117a) and *B. longibrachiatum* (Oudem.) Maire
    References: Hughes, 1953; Maire, 1903.

29. **Nematogonium** Desm. - Ann. Sci. nat., Bot., Sér. 2, *2*: 70 (1874)
    N. aurantiacum Desm. = *N. ferrugineum* (Pers. ex Fr.) Hughes
    = *Gonatorrhodiella* Thaxter - Bot. Gaz. *16*: 201 (1891)
      G. parasitica Thaxter = *N. parasiticum* (Thaxter) Hughes
    3 or 4 species
    References: Hughes, 1953; Barron, 1968.

. Gonatobotryum Sacc. - Syll. Fung. *4*: 278 (1886)
   G. fuscum Sacc.
2 or 3 species (Fig. 117d)
References: Hughes, 1953; Shigo, 1960; Kendrick et al., 1968.

. Cephaliophora Thaxter - Bot. Gaz. *35*: 153 (1903)
   C. tropica Thaxter (Fig. 118a)
= *Cephalomyces* Bain. - Bull. Soc. mycol. Fr. *23*: 109 (1907)
   C. nigricans Bain. = *C. irregularis* Thaxter
2 species
References: Goos, 1964.

. Oedocephalum Preuss - Linnaea *24*: 131 (1851)
   O. glomerulosum (Bull. ex Fr.) Sacc.
4 or 5 species (Fig. 118c)
ascigerous state: *Peziza* and other Pezizales (Webster et al., 1964; Berthet, 1964)
References: Tubaki, 1954.

. Amblyosporium Fres. - Beitr. Mykol. *3*: 99 (1863)
   A. botrytis Fres. = *A. spongiosum* (Pers. ex Fr.) Hughes (Fig. 118b)
1 or 2 species
References: Nicot and Durand, 1965; Pirozyński, 1969.

. Aureobasidium Viala & Boyer - Rev. gen. Bot. *3*: 369 (1891)
   A. vitis Viala & Boyer = *A. pullulans* (de Bary) Arn. (Fig. 118d)
= *Pullularia* Berkhout - Schimmelgesl. Monilia etc. p. 55 (1923)
   P. pullulans (de Bary) Berkh. = *A. pullulans*
ascigerous state: *Guignardia*. In pure culture a number of other ascomycetes have
the same characters, e.g. species of *Dothiora, Sydowia, Potebniamyces.*
2 or 3 species
References: von Arx, 1957, 1970.

. Blastobotrys Klopotek - Arch. Mikrobiol. *58*: 92 (1967)
   B. nivea Klopotek (Fig. 119a)
1 species.

. Sporothrix Hektoen & Perkins - J. exp. Med. *5*: 77 (1900)
   S. schenckii Hektoen & Perkins (Fig. 119b)
2 or 3 species, parasitic on man or saprophytic and mostly known as *Sporotrichum*. *Ophiostoma* species usually in pure culture form a *Sporothrix*-like conidial state.
References: Mariat et al., 1962; Emmons et al., 1963; Müller (1964, 1965) has
revised the human-pathogenic species described in the genus *Sporotrichum* Link.

. Dexhowardia Taylor - Mycopath. Mycol. appl. *40*: 306 (1970)
   D. tetraspora Taylor
1 species.

. Raffaelea v. Arx & Hennebert - Mycopath. Mycol. appl. *25*: 310 (1965)
   R. ambrosiae v. Arx & Hennebert (Fig. 119c)
5 or 6 species, all are ambrosia fungi of beetles (bark beetles)
References: Batra, 1967.

229

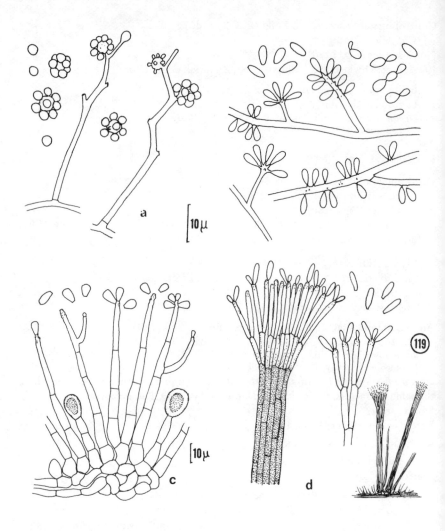

Fig. 119. a, *Blastobotrys nivea*; b, *Sporothrix schenckii*; c, *Raffaelea ambrosiae*; d, *Graphium ulmi* Schwarz (all orig.).

39. **Graphium** Corda - Icon. Fung. *1*: 18 (1937)
    G. penicillioides Corda
ascigerous states: *Ophiostoma, Petriella*
about 50 species are described, but the genus urgently needs revision.

40. **Pesotum** Crane & Schoknecht - Am. J. Bot. *60*: 347 (1973)
    P. ulmi (Schwarz) Crane & Schoknecht = *Graphium ulmi* Schwarz (Fig. 119)
ascigerous state: *Ophiostoma*

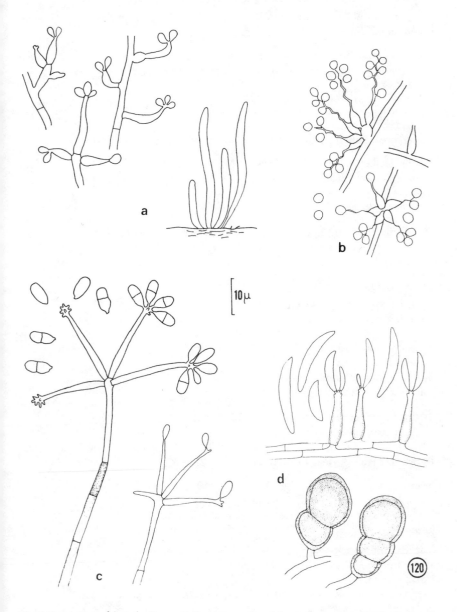

Fig. 120. a, *Isaria felina*; b, *Beauveria bassiana*; c, *Pseudobotrytis terrestris*; d, *Idriella lunata* (orig., d from W. Gams).

41. **Isaria** Fr. - Syst. mycol. *3*: 270 (1829)

    I. felina Fr. (= *I. cretacea* van Beyma) (lectotypus) (Fig. 120a)
about 10 species, mostly imperfectly known.
If *Isaria farinosa* Fr. were chosen as lectotype, as is proposed by Morris (1963),

the genus would be identical with *Paecilomyces*. In this species the conidia are borne on phialides.

42. **Pseudobotrytis** Krzemieniewska & Badura - Acta Soc. Bot. Polon. *23*: 727 (1954)
    P. fusca Krzemieniewska & Badura = *P. terrestris* (Timonin) Subram. (Fig. 120c)
    = *Umbellula* Morris - Mycologia *47*: 602 (1955)
      U. terrestris (Timonin) Morris = *P. terrestris*
    2 species
    References: Subramanian, 1956; Timonin, 1961; Litvinov, 1967.

43. **Costantinella** Matruchot - Rech. Dév. Mucéd. p. 97 (1892)
    C. cristata Matruchot = *C. terrestris* (Link ex S.F. Gray) Hughes
    2 species
    References: Ellis, 1971.

44. **Pseudogibellula** Samson & Evans - Acta bot. Neerl. *22*: 524 (1973)
    P. formicarum (Mains) Samson & Evans
    1 species.

45. **Idriella** Nelson & Wilhelm - Mycologia *48*: 547 (1956)
    I. lunata Nelson & Wilhelm (Fig. 120d)
    1 species.

46. **Microdochium** Syd. - Ann. mycol. *22*: 267 (1924)
    M. phragmitis Syd.
    5 or 6 species, often parasitic on Gramineae
    References: Sutton, Pirozynski and Deighton, 1972.

47. **Tritirachium** Limber - Mycologia *32*: 23 (1940)
    T. dependens Limber
    4 species (Fig. 122c)
    References: van Beyma, 1942; de Hoog, 1972, 1973.

48. **Beauveria** Vuill. - Bull. Soc. bot. Fr. *59*: 34 (1912)
    B. bassiana (Bals.) Vuill. (Fig. 120b)
    3 or 4 species, parasitic on insects or saprophytic in soil
    References: MacLeod, 1954, 1963; Benham and Miranda, 1953; de Hoog, 1972.

49. **Calcarisporium** Preuss - Linnaea *24*: 124 (1851)
    C. arbuscula Preuss (Fig. 121a)
    1 or 2 species
    References: Hughes, 1951; Watson, 1955; Tubaki, 1955; Barnett, 1958.

50. **Acrodontium** de Hoog - Stud. Mycol. *1*: 23 (1972)
    A. crateriforme (van Beyma) de Hoog
    ascigerous state: *Ascocorticium*
    8 species.

51. **Phaeoisaria** Höhnel - Sber. Akad. Wiss. Wien, Abt. 1,*118*: 329 (1909)
    P. bambusae Höhnel = *P. clematidis* (Fuck.) Hughes
    2 species.

2. **Dicyma** Boulanger - Rev. gén. Bot. *9*: 17 (1897)
   D. ampullifera Boulanger
   ascigerous state: *Ascotricha*.

3. **Hansfordia** Hughes - Mycol. Pap. *43*: 15 (1951)
   H. ovalispora Hughes
   about 10 species (Fig. 121b).

4. **Nodulisporium** Preuss in Klotzsch Herb. viv. mycol. no. 1272 (1849)
   N. ochraceum Preuss
   = *Acrostaphylus* Arn. - Bull. Soc. mycol. Fr. *69*: 272 (1953) sensu Subramanian (1956)
   A. hypoxyli Arn.
   ascigerous states: *Hypoxylon* Bull. ex Fr., *Nummularia* Tul. and related genera.
   about 20 species (Fig. 121d)
   References: Hughes, 1958; Meyer, 1959; Smith, 1962; Subramanian, 1956;
   Martin, 1967, 1968.

5. **Dactylaria** Sacc. - Michelia *2*: 20 (1830)
   D. purpurella Sacc. (Fig. 122a)
   = *Diplorhinotrichum* Höhnel - Sber. K. Akad. Wiss. Wien, math.-nat. Kl. *111*: 1040 (1902)
   D. candidulum Höhnel = *Dactylaria purpurella*
   = *Mirandina* Arn. - Bull. Soc. mycol. Fr. *68*: 205 (1952) (nomen nudum)
   M. corticola Arn.
   = *Pleurophragmium* Cost. - Les Mucédinées simples p. 100 (1888)
   P. bicolor Cost. = *D. parvispora* (Preuss) de Hoog & v. Arx
   about 20 species
   References: Hering, 1965; Roy and Gujarati, 1965; Bhatt and Kendrick, 1968.

6. **Ochroconis** de Hoog & v. Arx (in press)
   O. constricta (Abbott) de Hoog & v. Arx = *Scolecobasidium constrictum*
   Abbott (Fig. 122b)
   5 or 6 species.

7. **Monacrosporium** Oudem. - Ned. kruidk. Arch. 2,*4*: 250 (1885)
   M. elegans Oudem.
   about 25 species, mostly nemotode-trapping (Fig. 132b)
   References: Subramanian, 1963; Cooke and Dickinson, 1965.

8. **Dactylella** Grove - J. Bot., Lond. *22*: 195 (1884)
   D. minuta Grove
   = *Candelabrella* Rifai & R.C. Cooke - Trans. Br. mycol. Soc. *49*: 160 (1966)
   C. javanica Rifai & R.C. Cooke
   about 10 species, nematode-trapping
   References: see *Monacrosporium*.

9. **Arthrobotrys** Corda - Prachtflora p. 43 (1839)
   A. superba Corda
   about 10 species, common is *A. oligospora* Fres. (Fig. 122d)
   References: Drechsler, 1937; Meyer, 1959; Rifai and Cooke, 1966; Cooke and
   Godfrey, 1969; Haard, 1968.

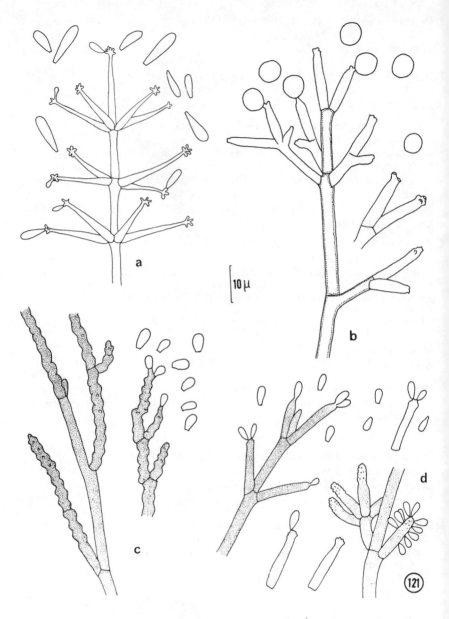

Fig. 121. a, *Calcarisporium arbuscula*; b, *Hansfordia pulvinata* (Berk. & Curt.) Hughes; c, *Geniculosporium serpens*; d, *Nodulisporium* spec. (orig., partly from W. Gams).

60. **Dematophora** Hartig - Unters. forstbot. Inst. München *3*: 95 (1883)

    D. necatrix Hartig, the conidial state of *Rosellinia necatrix* (Hartig) Berl. ex Prill.

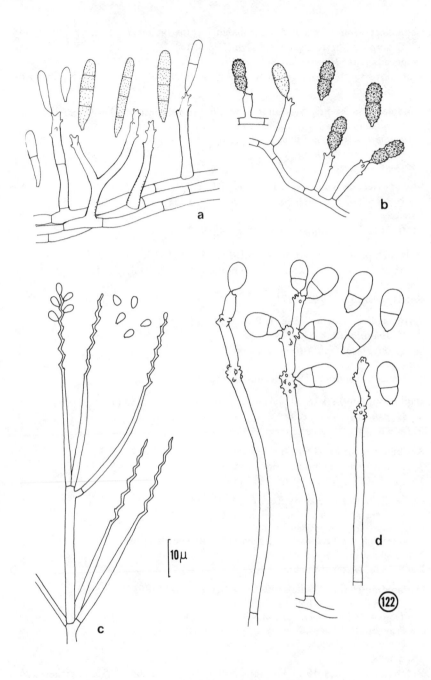

Fig. 122. a, *Dactylaria purpurella*; b, *Ochroconis constrictum* Abbot; c, *Tritirachium oryzae*; d, *Arthrobotrys oligospora* (a, b, c orig., d. from Domsch and Gams, 1970).

61. **Geniculosporium** Chesters & Greenhalgh - Trans Br. mycol. Soc. *47*: 393 (1964)
     G. serpens Chesters & Greenhalgh (Fig. 121c)
     ascigerous state: *Hypoxylon* Bull. ex Fr.
     2 or more species. The genus is very close to *Nodulisporium* and hardly distin-
     guishable from it.

62. **Conoplea** Pers. ex Fr. - Syst. mycol. *3*: 490 (1832) sensu Hughes (1960)
     C. sphaerica Pers. = *C. olivacea* Fr.
     = *Streptothrix* Corda - Prachtflora p. 27 (1839)
       S. fusca Corda = *C. fusca* Pers.
     6 species.

63. **Virgaria** Nees ex Sacc. - Syll. Fung. *4*: 281 (1886)
     V. nigra (Link ex Fr.) Nees ex Sacc. (Fig. 123a)
     1 species
     References: Subramanian, 1956; Hughes, 1953.

64. **Melanographium** Sacc. - Ann. mycol. *11*: 557 (1913)
     M. spleniosporum Sacc. = *M. selenioides* (Sacc. & Paoletti) M.B. Ellis
     5 species
     References: Ellis, 1963, 1971.

65. **Verticicladiella** Hughes - Can. J. Bot. *31*: 653 (1953)
     V. abietina (Peck) Hughes
     ascigerous state: *Ophiostoma, Europhium*
     9 species (Fig. 123b)
     References: Kendrick, 1962; Kendrick and Molnar, 1965.

66. **Haplographium** Berk. & Br. - Ann. Mag. nat. Hist. 3, *3*: 360 (1859)
     H. delicatum Berk. & Br.
     References: Hughes, 1953 (Fig. 123c).

67. **Arxiella** Papendorf - Trans. Br. mycol. Soc. *50*: 73 (1967)
     A. terrestris Papendorf (Fig. 123d)
     1 species.

68. **Gyoerffiella** Kol - Folia cryptog. *1*: 618 (1928)
     G. tatrica Kol = *G. rotula* (Höhnel) Marvanová
     = *Ingoldia* Petersen - Mycologia *54*: 147 (1962)
       I. craginiformis Petersen = *G. craginiformis* (Petersen) Marvanová
     4 or 5 species
     References: Marvanová et al., 1967.

69. **Dicranidion** Harkn. - Bull. Calif. Acad. Sci. *1*: 163 (1885)
     D. fragile Harkn. (Fig. 123f)
     = *Pedilospora* Höhnel - Sber. K. Akad. Wiss. Wien, math.-nat. Kl. *111*: 1047 (1902)
       P. parasitans Höhnel = *D. fragile*
     ascigerous state: *Orbilia* Fr.
     1 species
     References: Hughes, 1953; Peek and Solheim, 1958; Tubaki, 1958; Berthet, 1964.

70. **Scolecobasidium** Abbott - Mycologia *19*: 29 (1927)
     S. terreum Abbott
     2 species. Some species described as *Scolecobasidium* have to be classified in
     Ochroconis.

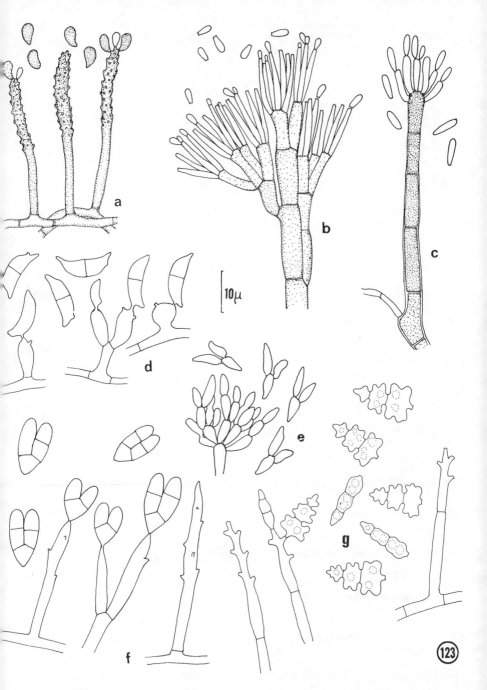

Fig. 123. a, *Virgaria nigra*; b, *Verticicladiella procera* Kendrick; c, *Haplographium* spec.; d, *Arxiella terrestris*; e, *Tricellula aquatica* Webster; f, *Dicranidion fragile*; g, *Dendrosporium lobatum* (a, d orig., b, c, e orig. from W. Gams, f from Tubaki, 1958).

71. **Dendrosporium** Plakidas & Edgerton - Mycologia *28*: 83 (1936)
    D. lobatum Plakidas & Edgerton (Fig. 123g)
    References: Crane, 1972.

72. **Tricellula** van Beverwijk - Antonie van Leeuwenhoek *20*: 15 (1954)
    T. inaequalis van Beverwijk
= *Volucrispora* Haskins - Can. J. Microbiol. *4*: 278 (1958)
    V. aurantiaca Haskins = *Tricellula aurantiaca* (Haskins) v. Arx
    5 species (Fig. 123e)
    References: Petersen, 1962; Webster, 1959; Ingold et al., 1968.

73. **Nakataea** Hara - Diseases of the rice plant 2, p. 185 (1939)
    N. sigmoidea Hara
= *Vakrabeeja* Subram. - J. Ind. bot. Soc. *35*: 465 (1956)
    V. sigmoidea (Cav.) Subram.
    ascigerous state: *Magnaporthe* Krause & Webster
    2 species
    References: Matsushima, 1971 (as *Vakrabeeja*); Krause & Webster (1972).

74. **Pseudofusarium** Matsushima - Microfungi Solomon Isl. and Papua-New Guinea p. 46 (1971)
    P. fusarioideum Matsushima
    2 species.

75. **Camposporium** Harkness - Bull. Calif. Acad. Sci. *1*: 37 (1884)
    C. antennatum Harkness
    3 species
    References: Hughes, 1951.

76. **Subulispora** Tubaki - Trans. mycol. Soc. Japan *12*: 20 (1971)
    S. procurvata Tubaki
    1 species.

77. **Speiropsis** Tubaki - J. Hattori Bot. Lab. *20*: 171 (1958)
    S. pedatospora Tubaki
    2 species
    References: Matsushima, 1971.

78. **Isthmolongispora** Matsushima - Microfungi Solomon Isl. and Papua-New Guinea p. 32 (1971)
    I. intermedia Matsushima
    2 species.

79. **Wiesneriomyces** Koorders - Verh. Akad. Amsterd. 2,*13*: 246 (1907)
    W. javanicus Koordeï s
    References: Subramanian, 1956; Matsushima, 1971.

80. **Beltrania** Penzig - Nuovo G. bot. Ital. *14*: 72 (1882)
    B. rhombica Penzig (Fig. 124e)
    about 5 species
    References: Tubaki, 1958; Pirozynski, 1963.

1. **Beltraniella** Subram. - Proc. Ind. Acad. Sci. *36*: 227 (1952)
   B. odinae Subram.
   References: Tubaki, 1958. For related genera see Pirozynski, 1963; Ellis, 1971.

2. **Centrospora** Neergaard - Zentbl. Bakt. Parasitkde, Abt. 2,*104*: 410 (1942)
   C. acerina (Hartig) Newhall (Fig. 132 f)
   3 or 4 species
   References: Newhall, 1946; Petersen, 1962; Nilsson, 1964.

3. **Mastigosporium** Riess in Fres. - Beitr. Mykol. p. 56 (1852)
   M. album Riess
   4 species, parasitic on grasses
   References: Sprague, 1950; Hughes, 1951; Austwick, 1954; Bollard 1950; Schlösser, 1970; Gunnerbeck, 1971.

4. **Pleiochaeta** Hughes - Mycol. Pap. *36*: 39 (1951)
   P. setosa (Kirchner) Hughes
   second species: *P. albizziae* (Petch) Hughes.

5. **Pyricularia** Sacc. - Michelia *2*: 20 (1880)
   P. grisea (Cooke) Sacc.
   ascigerous state: *Massarina* Sacc.
   4 or 5 species; *P. oryzae* Cavara is parasitic on rice. The genus is closely related to *Ovularia* and *Ramularia.*
   References: MacGarvie, 1968; Luttrell, 1954; Webster, 1965.

6. **Hadrotrichum** Fuckel - Fungi Rhenani no. 1522 (1865)
   H. phragmitis Fuckel
   1 species
   References: Hughes, 1953.

7. **Polythrincium** Kunze ex Fr. - Syst. mycol. *3*: 367 (1832)
   P. trifolii Kunze ex Fr.
   1 species, parasitic on leaves of *Trifolium*, with *Mycosphaerella killianii* Petr. (= *Cymadothea trifolii* Wolf) as its ascigerous state.
   References: Wolf, 1935; Petrak, 1941; Hughes, 1953.

8. **Veronaea** Cif. & Montem. - Atti. Ist. bot. Lab. critt. Pavia 5, *15*: 68 (1957)
   V. botryosa Cif. & Montem.
   = *Sympodina* Subram. & Lodha - Antonie van Leeuwenhoek *30*: 317 (1964)
   S. coprophila Subram. & Lodha
   2 or 3 species (Fig. 124a)
   References: Papendorf, 1969.

9. **Cordana** Preuss - Linnaea *24*: 129 (1851)
   C. pauciseptata Preuss (Fig. 124c)
   2 or 3 species
   References: Hughes, 1955.

10. **Fusicladium** Bon. - Handb. allg. Mykol. p. 80 (1851)
   F. virescens Bon. = *F. pyrorum* (Lib.) Fuckel (Fig. 124b)

= *Megacladosporium* Viennot-Bourgin - Les Champignons parasites des plantes cultivées *1*: 489 (1949)

    M. pyrorum (Lib.) Viennot-Bourgin = *F. pyrorum*

ascigerous state: *Venturia*

about 10 species, parasitic on higher plants. The genus is closely related to *Spilocaea*.

References: Hughes, 1953.

91. **Passalora** Fr. - Summa Veg. Scand. *2*: 500 (1849)

    P. bacilligera (Mont. & Fr.) Mont. & Fr.

ascigerous state: *Mycosphaerella*

about 10 species, causing leaf spots, mostly sterile in pure culture

References: Hughes, 1953; Ellis et al., 1951.

92. **Cercosporella** Sacc. - Michelia *2*: 20 (1880)

    C. persica (Sacc.) Sacc. or *C. cana* (Sacc.) Sacc.

ascigerous state: *Mycosphaerella*

about 50 species, causing leaf spots, mostly sterile in pure culture (Fig. 124d)

References: Sprague, 1937, 1950.

93. **Cercospora** Fres. - Beitr. Mykol. *3*: 91 (1863)

    C. apii Fres.

ascigerous state: *Mycosphaerella*

More than 1000 species are described as causal agents of leaf spots. Most of them in pure culture develop only a dark, slow growing mycelium and no conidia.

References: Chupp, 1959; Deighton, 1959.

94. **Cercosporidium** Earle - Muehlenbergia *1*(2): 16 (1901)

    C. chaetomium (Cooke) Deighton

ascigerous state: *Mycosphaerella*

about 17 species

References: Deighton, 1967.

95. **Phaeoisariopsis** Ferraris - Ann. mycol. 7: 280 (1909)

    P. griseola (Sacc.) Ferraris, parasitic on beans.

96. **Brachysporium** Sacc. - Syll. Fung. *4*: 427 (1886)

    B. obovatum (Berk.) Sacc.

3 species

References: Hughes, 1951, 1955, 1958; Ellis, 1966.

97. **Cacumisporium** Preuss - Linnaea *24*: 130 (1851)

    C. capitatulum (Corda) Hughes

References: Goos, 1969.

98. **Periconiella** Sacc. - Atti Ist. ven. Sci., ser. 6,*3*: 727 (1885)

    P. velutina (Wint.) Sacc.

about 20 species

References: Ellis, 1967, 1971.

99. **Deightoniella** Hughes - Mycol. Pap. *48*: 27 (1952)

    D. africana Hughes

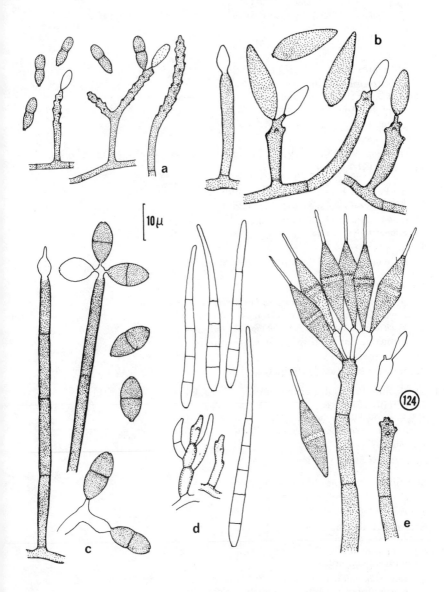

Fig. 124. a, *Veronaea simplex* Papendorf; b, *Fusicladium pyrorum*; c, *Cordana pauciseptata*; d, *Cercosporella herpotrichoides* Fron; e, *Beltrania rhombica* (orig., d from W. Gams, c from A. C. Stolk).

8 species
References: Ellis, 1971.

Corynespora Güssow - Z. Pfl. Krankh., Pfl. Path., Pfl. Schutz *16*: 13 (1906)
    C. mazei Güssow = *C. cassiicola* (Berk. & Curt.) Wei

= *Cuspidosporium* Cif. - Sydowia *9*: 303 (1955)
   C. cuspidatum (Sacc.) Cif. = *Corynespora pulviniformis* (Sacc.) Hughes
about 20 species
References: Ellis, 1957, 1961; Hughes, 1958.

101. **Stemphylium** Wallr. - Fl. crypt. Germ. *2*: 300 (1833)
   S. botryosum Wallr. (Fig. 125b)
= *Fusicladiopsis* Maire - Bull. Soc. bot. Fr. *53*: 187 (1906)
   F. conviva Maire = *S. botryosum*
ascigerous state: *Pleospora*
5 or 6 species
References: Wiltshire, 1938; Neergaard, 1945; Groves and Skolko, 1944; Simmons, 1967, 1969.

102. **Torula** Pers. ex Fr. - Syst. mycol. *3*: 499 (1832)
   T. herbarum Pers. ex Fr. (Fig. 125a)
= *Satwalekera* Rao & al. - Nova Hedwigia *18*: 637 (1969)
   S. sundara Rao & Rao
4 or 5 species
References: Tubaki, 1958; Joly, 1964; Misra, 1967.

103. **Alternaria** Nees ex Fr. - Syst. mycol. *1*: 46 (1821)
   A. tenuis Nees = *A. alternata* (Fr.) Keissler (Fig. 125c)
= *Macrosporium* Fr. - Syst. mycol. *3*: 374 (1832)
   M. cheiranthi (Lib.) Fr. = *A. cheiranthi* (Lib.) Bolle
ascigerous states: *Leptosphaeria, Clathrospora*
about 50 species
References: Neergaard, 1945; Joly, 1964; Simmons, 1952, 1967; Rao, 1965.

104. **Ulocladium** Preuss in Sturm Deutschl. Flora, Pilze 3, *3*: 83 (1851)
   U. botrytis Preuss
= *Pseudostemphylium* Subram. - Curr. Sci *30*: 423 (1961)
   P. consortiale (Thüm.) Subram. = *U. consortiale* (Thüm.) Simmons (Fig. 125d)
9 species
References: Simmons, 1967.

105. **Embellisia** Simmons - Mycologia *63*: 380 (1971)
   E. allii (Campanile) Simmons
3 species
References: de Hoog and Muller, 1973.

106. **Dendryphion** Wallr. - Fl. crypt. Germ. *2*: 300 (1833)
   D. comosum Wallr.
= *Brachycladium* Corda - Icon. Fung. *2*: 14 (1838)
   B. penicillatum Corda = *D. penicillatum* (Corda) Fr.
= *Entomyclium* Wallr. - Fl. crypt. Germ. *2*: 189 (1833)
   E. folliculatum Wallr. = *D. nanum* (Nees ex Fr.) Hughes (Fig. 116a)
5 or 6 species
References: Hughes, 1953, 1958; Reisinger, 1968; Shoemaker, 1968.

107. **Dendryphiella** Bubák & Ranoj. - Ann. mycol. *12*: 417 (1914)
   D. vinosa (Berk. & Curt.) Reisinger

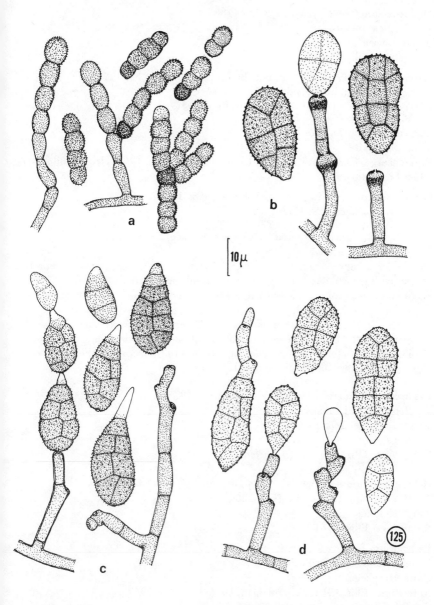

Fig. 125. a, *Torula herbarum*; b, *Stemphylium botryosum*; c, *Alternaria alternata*; d, *Ulocladium consortiale* (orig.).

2 species
References: Pugh and Nicot, 1964; Reisinger, 1968.

3. **Dendryphiopsis** Hughes - Can. J. Bot. *31*: 655 (1953)
    D. atra (Corda) Hughes.

243

109. **Diplococcium** Grove - J. Bot. *23*: 167 (1885)
    D. spicatum Grove
    3 or 4 species
    References: Ellis, 1963.

110. **Curvularia** Boedijn - Bull. Jard. bot. Buitenzorg 3, *13*, 1: 127 (1933)
    C. lunata (Wakker) Boedijn (Fig. 126a)
    = *Malustela* Batista & Lima - Publ. Inst. Micol. Recife *263*: 7 (1960)
    M. aeria Batista & al. = *C. lunata*
    ascigerous state: *Cochliobolus*
    about 30 species
    References: Ellis, 1966; Kamat and Rao, 1970; Nelson, 1964; Nelson and Haasis, 1964; Corbetta, 1965.

111. **Drechslera** Ito - Proc. Imp. Acad. Tokyo *6*: 455 (1930)
    D. tritici-vulgaris (Nisikado) Ito
    = *Bipolaris* Shoemaker - Can. J. Bot. *37*: 879 (1959)
    B. maydis (Nisikado) Shoemaker = *D. maydis* (Nisikado) Subram. & Jain
    ascigerous states: *Pyrenophora, Cochliobolus*
    Drechslera comprises a group of mostly graminicolous *Helminthosporium*-like species with geniculate conidiophores. About 60 species are described.
    A common saprophyte of warmer areas is *Drechslera spicifera* (Bain.) v. Arx (Fig. 126c)
    The genus *Bipolaris* has been erected by Shoemaker (1959) for species with fusiform conidia, but has been rejected by Subramanian and Jain (1966). Lutrell (1963, 1964) preferred to use the name *Helminthosporium* for the whole complex (and see Ammon, 1963; Rapilly, 1964, 1966).

112. **Exserohilum** Leonard & Suggs - Mycologia *65* (1973)
    E. prolatum Leonard & Suggs
    ascigerous state: *Setosphaeria*
    8 species
    References: Ellis, 1971 (as *Drechslera*).

113. **Helminthosporium** Link ex Fr. - Syst. mycol. *3*: 359 (1832)
    H. velutinum Link ex Fr. (Fig. 126d)
    about 15 species
    References: Ellis, 1961.

114. **Exosporium** Link ex Schlecht. - Flora Berol. *2*: 140 (1824)
    E. tiliae Link ex Schlecht.
    about 10 species
    References: Ellis, 1961; Subramanian, 1956; Carrama, 1967.

115. **Dichotomophthora** Mehrlich & Fitzp. - Mycologia *27*: 543 (1935)
    D. portulacae Mehrlich & Fitzp.
    2 species
    References: Rao, 1966.

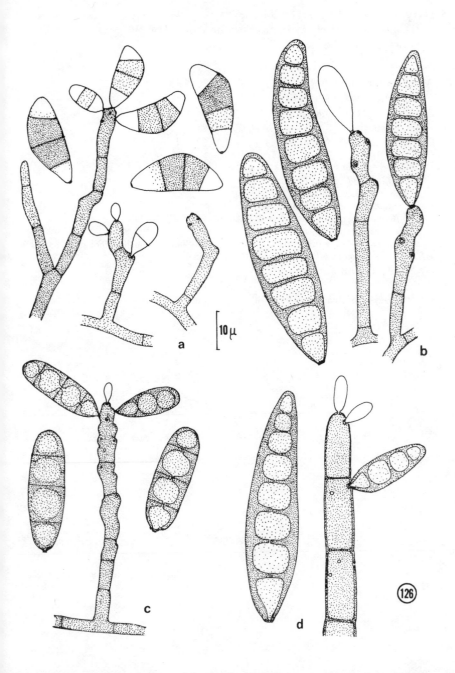

Fig. 126. a, *Curvularia lunata*; b, *Drechslera (Helminthosporium) victoriae* (Meehan & Murphy) Subram. & Jain; c, *Drechslera spicifera* (Bain.) v. Arx; d, *Helminthosporium velutinum* (orig., d from a herbarium specimen).

116. **Dictyoarthrinium** Hughes - Mycol. Pap. *48*: 29 (1952)
　　D. quadratum Hughes = *D. sacchari* (Stevenson) Damon
　1 species
　References: Ellis, 1971.

117. **Arthrinium** Kunze ex Fr. - Syst. mycol. *3*: 376 (1832)
　　A. caricicola Kunze ex Fr.
　ascigerous state: *Physalospora*
　about 18 species
　References: Ellis, 1965.

118. **Papularia** Fr. - Summa Veg. Scand. p. 509 (1849)
　　P. arundinis (Corda) Fr.
　= *Coniosporium* Sacc. - Michelia *2*: 21 (1880)
　　C. arundinis (Corda) Sacc. = *P. arundinis*
　= *Innatospora* van Beyma - Verh. K. ned. Akad. Wet. 2,*26*,4:5 (1929)
　　I. rosea van Beyma = *C. arundinis*
　= *Pseudobasidium* Tengwall - Meded. phytopath. Lab. Willie Commelin Scholten *6*: 38 (1924)
　　P. bicolor Tengwall = *P. phaeosperma* (Pers.) Höhnel
　ascigerous state: *Apiospora*
　2 species (Fig. 116d)
　References: Subramanian, 1971; Ellis, 1965 (as *Arthrinium*).

Moniliales Key V (genera with solitary aleurioconidia or chlamydospores)

1. Conidia 1-celled or composed of 2 or 3 unequal cells, conidiogenous cells usually not annellated 2
1. Conidia 2- or usually multicelled, cells equal or nearly so 31
2. Conidia hyaline.. 3
2. Conidia pigmented... 12
3. Ambrosia-fungi, conidiogenous cells usually with constrictions, sporodochia usually present, conidia terminal, spherical ....*Ambrosiella* (1)
3. Conidiogenous cells without constrictions, no ambrosia-fungi 4
4. Conidiophores ending in short spines, sporodochia usually present, conidia often lateral, colonies reddish or orange .... *Botryoderma* (12)
4. Not combining above characters 5
5. Conidia usually less than 10 μm in diameter 6
5. Conidia larger 7
6. Conidia roundish in outline, usually borne laterally on wide hyphae ....
.... *Beniowskia* (25)
    (Hyphae narrow, conidia lateral, small: cf. *Trichosporiella* (25a))
6. Conidia truncate at base, often thick-walled or verrucose, hyphae narrow ...
.... *Chrysosporium* (2)
    (Conidia thin-walled: cf. *Scedosporium*)
7. Dermatophytes (isolated from man or animals), phialoconidia absent 8
7. Not dermatophytic, often fungicolous, phialoconidia often present 9
8. Conidia warty, usually formed singly, blastoconidia absent. *Histoplasma* (26)
8. Conidia usually smooth, blastoconidia often also present.. *Blastomyces* (27)
9. Fungicolous (usually isolated from Basidiomycetes or Discomycetes) 10
9. Not fungicolous 12
10. Conidia inequally 2-celled, with a small basal cell...........*Mycogone* (5)
10. Conidia 1-celled 11
11. Conidia with bullate processes .... *Stephanoma* (4)
11. Conidia without bullate processes .... *Sepedonium* (3)
12. Conidia composed of 2-3 unequal cells........................... 13
12. Conidia 1-celled 15
13. Conidia 2-celled, basal cell small .... *Chlamydomyces* (6)
13. Conidia 3-celled, 2 small cells are laterally attached to a larger cell 14
14. Conidia large, verrucose, conidiophores absent............*Acrospeira* (8)
14. Conidia small, borne on verticillately arranged, tapering cells ....
.... *Physalidium* (43)

28. Conidia smooth or verrucose 29

29. Conidia obclavate or obpyriform, verrucose .... *Echinobotryum* (19)

29. Conidia spherical, clavate or ellipsoidal 30

30. Phialoconidia absent or small, ellipsoidal or spherical .... *Humicola* (13)

30. Phialoconidia cylindrical, borne in tube-like phialides ...... cf. *Chalaropsis*

31. Conidia hyaline, with transverse septa only (dermatophytes) 32

31. Conidia pigmented or muriform when hyaline. 35

32. Macroconidia clavate, cylindrical or obovate, multiseptate 33

32. Macroconidia fusiform, often verrucose 34

33. One-celled microconidia with a truncate base (*Chrysosporium*) usually also present .... *Trichophyton* (30)

33. Microconidia usually absent. .... *Epidermophyton* (31)

34. *Chrysosporium*-microconidia usually also present .... *Microsporon* (28)

34. Microconidia usually absent (keratinophilic) .... *Keratinomyces* (29)

35. Conidia when detached with remains of the narrow cylindrical conidiogenous cell .... *Pithomyces* (39)

35. Conidia when detached without remains of the conidiogenous cell...... 36

36. Synnemata present, conidia large, muriform .... *Kostermansinda* (41)

36. Synnemata absent, sporodochia may be present 37

37. Conidiogenous cells with a swollen tip, percurrent, becoming nodose 38

37. Conidiogenous cells not nodose, not percurrent 39

38. Conidia with transverse septa only .... cf. *Deightoniella*

38. Conidia muriform .... .cf. *Stemphylium*

39. Conidia 2-3-celled, obpyriform or with a barrel-shaped central cell, conidiogenous cells short, colonies restricted (parasitic on plants) 40

39. Conidia usually multiseptate or pseudoseptate 41

40. Conidia obpyriform or obclavate, usually 2-celled .... cf. *Spilocaea*

40. Conidia 2-3-celled, central cell barrel-shaped .... cf. *Pollaccia*

41. Conidia muriform, with several appendages .... cf. *Petrakia*

41. Conidia without appendages when muriform 42

42. Conidia roundish in outline, usually verrucose, borne in black sporodochia ..
.... cf. *Epicoccum*
(Conidia hyaline or yellowish, chlamydospore-like: cf. *Diheterospora*)

42. Conidia usually not roundish in outline 43

43. Conidia fusiform, pseudoseptate, borne on narrow conidiogenous cells .....
.... *Murogenella* (37)

List of genera

1. **Ambrosiella** Brader ex v. Arx & Hennebert - Mycopath. Mycol. appl. *25*: 313 (1965)
   A. xylebori Brader (Fig. 127a)
   8 species, all are ambrosia fungi of bark beetles
   References: Batra, 1967.

2. **Chrysosporium** Corda in Sturm Deutschl. Fl., Pilze 3, *3*, 13: 85 (1833)
   Ch. corii Corda = *Ch. merdarium* (Link ex Fr.) Carmichael (Fig. 127b)
   = *Aleurisma* auct.
   = *Geomyces* Traaen - Nyt Mag. Naturvid. *5 2*: 28 (1914)
   G. vulgaris Traaen = *Ch. pannorum* (Link) Hughes
   = *Glenosporella* Nann. - Atti Ist. bot. Univ. Pavia 4, 2: 98 (1930)
   G. albicans (Nieuwenh.) Nann. = *Ch. pannorum*
   ascigerous states: *Gymnoascus, Arthroderma, Ctenomyces, Anixiopsis, Aphano-ascus* and other genera
   about 15 species, and compare *Blastomyces*
   References: Carmichael, 1962; Pitt, 1966.

3. **Sepedonium** Link ex Fr. - Syst. mycol. *3*: 428 (1832)
   S. chrysospermum (Bull.) Link ex Fr. (Fig. 127c)
   ascigerous states: *Hypomyces (Apiocrea), Corynascus*
   4 or 5 species
   References: Damon, 1952; Tubaki, 1955.

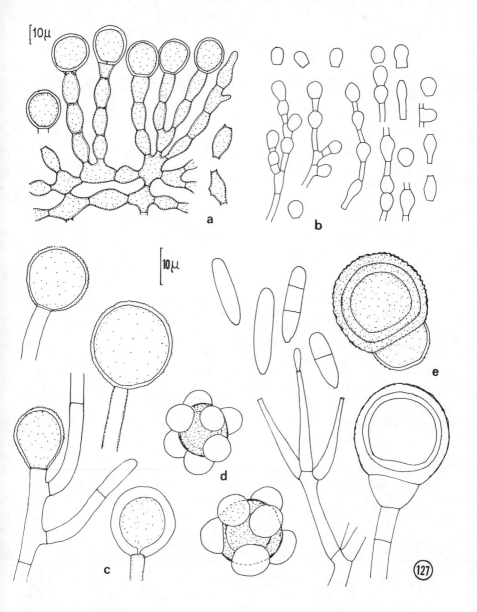

Fig. 127. a, *Ambrosiella xylebori*; b, *Chrysosporium merdarium*; c, *Sepedonium chrysospermum*; d, *Stephanoma strigosum*; e, *Mycogone rosea* (a orig., b from Domsch and Gams, 1970, c, d, e orig. from W. Gams).

4. **Stephanoma** Wallr. - Fl. crypt. Germ. *2*: 269 (1833)
    S. strigosum Wallr. (Fig. 127d)
2 species, parasitic on Discomycetes
References: Tubaki, 1963; van Zinderen-Bakker, 1934.

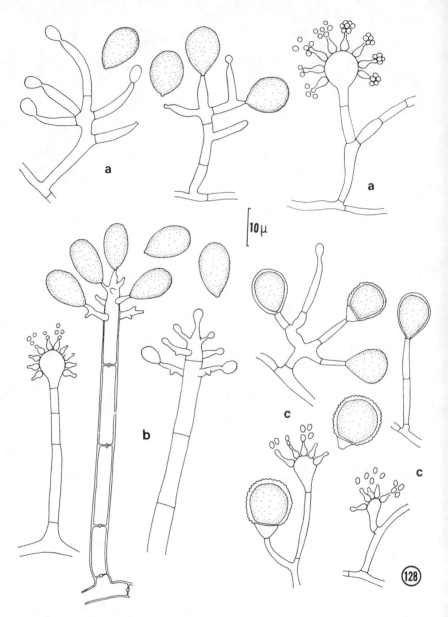

Fig. 128. a, *Acremoniella atra*; b, *Olpitrichum macrosporum*; c, *Chlamydomyces palmarum* (orig.).

5. **Mycogone** Link - Linn. Spec. Plant. *1*: 30 (1824)
   M. rosea Link ex Link (Fig. 127e)
   = *Coccosporella* Karst. - Acta Soc. Fauna Fl. Fenn. *9*: 11 (1893)
     C. calospora Karst. = *M. calospora* (Karst.) Höhnel

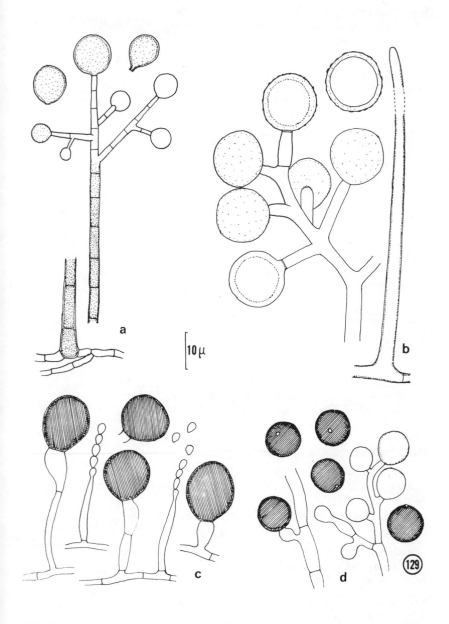

Fig. 129. a, *Staphylotrichum coccosporum*; b, *Botryotrichum piluliferum*; c, *Humicola grisea*; d, *Gilmaniella humicola* (orig., d from A. C. Stolk).

ascigerous state: *Hypomyces*
5 or 6 species, parasitic on Basidiomycetes (mushrooms)
References: Howell, 1939; Tubaki, 1955.

6. **Chlamydomyces** Bain. - Bull. Soc. mycol. Fr. *28*: 238 (1907)
   Ch. diffusus Bain. = *Ch. palmarum* (Cooke) Mason (Fig. 128d)
   References: Mason, 1933.

7. **Botryotrichum** Sacc. & March. - Bull. Soc. R. Bot. Belg. *24*: 66 (1885)
   B. piluliferum Sacc. & March. (Fig. 129b)
   = *Coccospora* Goddard - Bot. Gaz. *56*: 264 (1913)
      C. agricola Goddard = *B. piluliferum*
   ascigerous state: *Chaetomium piluliferum* Daniels (1961)
   2 species
   References: Downing, 1953; Blochwitz, 1914.

8. **Acrospeira** Berk. & Br. - Introd. crypt. Bot. p. 305 (1857)
   A. mirabilis Berk. & Br.
   = *Spirospora* Mang. & Vinc. - Bull. Soc. mycol. Fr. *36*: 96 (1920)
      S. castaneae Mang. & Vinc. = *A. mirabilis*
   1 species, closely related to *Chlamydomyces*.

9. **Acremoniella** Sacc. - Syll. Fung. *4*: 302 (1886)
   A. atra Sacc. (Fig. 128a)
   = *Harzia* Cost. - Les Mucédinées simples p. 42 (1888)
      H. acremonioides (Harz) Cost. = *A. atra*   -
   ≡ *Eidamia* Lindau in Rabenh. Krypt. Fl. *1, 8*: 123 (1907)
   = *Monopodium* Delacr. - Bull. Soc. mycol. Fr. *6*: 99 (1890)
      M. uredopsis Delacr. = *A. atra*
   2 or 3 species
   References: Mason, 1933; Tubaki, 1963; Nicot, 1962; Focke, 1962; Onions, 1968.

10. **Staphylotrichum** Meyer & Nicot - Bull. Soc. mycol. Fr. *72*: 322 (1956)
    S. coccosporum Meyer & Nicot (Fig. 129a)
    = *Botrydiella* Badura - Allionia *9*: 182 (1963)
       B. bicolor Badura = *S. coccosporum*
    1 species
    References: Maciejowska and Williams, 1963.

11. **Olpitrichum** Atk. - Bot. Gaz. *19*: 244 (1849)
    O. carpophilum Atk. = *O. macrosporum* (Farl.) Sumst. (Fig. 128b)
    2 or 3 species
    References: Linder, 1942 (as *Oidium*). The genus is closely related to *Acremoniella*.

12. **Botryoderma** Papendorf & Upadhyay - Trans. Br. mycol. Soc. *52*: 257 (1969)
    B. lateritium Papendorf & Upadhyay
    2 species from soil.

13. **Humicola** Traaen - Nyt Mag. Naturvid. *32*: 20 (1914)
    H. fuscoatra Traaen
    = *Melanogone* Wollenw. & Richter - Zentbl. Bakt. Parasitkde, Abt. 2, *89*: 74 (1934)
       M. puccinioides Wollenw. & Richter = *H. grisea* Traaen
    = *Monotospora* auct.
    8 species (Fig. 129c)
    References: Fassatiová, 1967; Gilman, 1957; Omvik, 1955; White and Dowing, 1953; Fergus, 1964.

14. **Acrogenospora** M.B. Ellis - Demat. Hyphomycetes p. 114 (1971)
    A. sphaerocephala (Berk. & Br.) M.B. Ellis
    ascigerous state: *Farlowiella*
    2 species.

15. **Thermomyces** Tsiklinsky - Ann. Inst. Pasteur *13*: 500 (1899)
    T. lanuginosus Tsiklinsky
    3 or 4 species
    References: Pugh et al., 1964; Apinis and Eggins, 1966.

16. **Gilmaniella** Barron - Mycologia *56*: 514 (1964)
    G. humicola Barron (Fig. 129d)
    = *Adhogamina* Subram. & Lodha - Antonie van Leeuwenhoek *30*: 328 (1964)
    A. ruchira Subram. & Lodha = *G. humicola*
    1 species.

17. **Wardomyces** Brooks & Hansf. - Trans. Br. mycol. Soc. *8*: 137 (1922)
    W. anomalus Brooks & Hansf. (Fig. 130a)
    4 or 5 species
    Gymnodochium Massee & Salmon seems to be an older name.
    References: Hennebert, 1962, 1968; Gams, 1968; Dickinson, 1964, 1966; Sugiyama et al., 1968.

18. **Mammaria** Ces. - Flora *12*: 207 (1854)
    M. echinobotryoides Ces. (Fig. 130b)
    1 species
    References: Hughes, 1957; Hennebert, 1968; Sugiyama, 1969.

19. **Echinobotryum** Corda in Sturm Deutschl. Flora, Pilze, 3, *12*: 51 (1831)
    E. atrum Corda (is the aleuriosporic state of *Cephalotrichum stemonitis* (Pers.) Link ex Fr.)
    1 species
    References: Hennebert, 1968.

20. **Nigrospora** Zimmerm. - Zentbl. Bakt. Parasitkde, Abt. 2, *8*: 220 (1902)
    N. panici Zimmerm.
    = *Dichotomella* Sacc. - Ann. mycol. *12*: 312 (1914)
    D. areolata Sacc. = *N. oryzae* (Berk. & Br.) Petch
    = *Basisporium* Molliard - Bull. Soc. mycol. Fr. *18*: 170 (1902)
    B. gallarum Molliard = *N. oryzae*
    ascigerous state: *Khuskia* Hudson (= *Scirrhiella* Speg. = *Apiospora* Sacc.)
    4 or 5 species (Fig. 130c)
    References: Mason, 1933; Standon, 1943; Hudson, 1963.

21. **Zygosporium** Mont. - Ann. Sci. nat., Sér. 2, *17*: 120 (1842)
    Z. oscheoides Mont.
    synonyms: see Ellis, 1971
    about 10 species (fig. 130d).

22. **Asteromyces** F. & Mme Moreau ex Hennebert - Can. J. Bot. *40*: 1211 (1962)
    A. cruciatus F. & Mme Moreau ex Hennebert
    1 species.

23. **Allescheriella** P. Henn. - Hedwigia *36*: 244 (1897)
   A. crocea (Mont.) Hughes
   1 species
   References: Hughes, 1951.

24. **Scedosporium** Sacc. - Ann. mycol. *9*: 254 (1911)
   S. apiospermum (Sacc.) Sacc., the conidial state of *Petriellidium boydii*. The genus *Scedosporium* is not validely published.

25. **Beniowskia** Rac. - Paras. Algen Pilze Java's *2*: 37 (1900)
   B. graminis Rac. = *B. sphaeroidea* (Kalchbr. & Cooke) Mason
   2 or 3 species.

25a. **Trichosporiella** Kamyschko ex W. Gams & Domsch - Nova Hedwigia *18*: 19 (1969)
   T. hyalina Kamyschko = *T. cerebriforme* (de Vries & Kleine-Natrop) W. Gams
   1 species.

26. **Histoplasma** Darling - J. Am. med. Ass. *46*: 1283 (1906)
   H. capsulatum Darling (Fig. 131c)
   3 species
   References: Dodge, 1935; Emmons et al., 1963.

27. **Blastomyces** Cost. & Roll. - Bull. Soc. mycol. Fr. *4*: 153 (1888)
   B. luteus Cost. & Roll. = *Chrysosporium merdarium* (Link) Carmichael
   = *Zymonema* Beurm. & Gougerot - Tribune méd. *42*: 503 (1909)
     Z. gilchristi Beurm. & Gougerot = *B. dermatitidis* Gilchrist & Stokes
   = *Blastomycoides* Castellani - Am. J. trop. Med. *8*: 381 (1928)
     B. dermatitidis (Gilchrist & Stokes) Castellani = *B. dermatitidis*
   = *Gilchristia* Red. & Cif. - J. Trop. Med. Hyg. *37*: 280 (1934)
     G. dermatitidis (Gilchrist & Stokes) Redaelli & Ciferri = *B. dermatitidis* (= *Chrysosporium dermatitidis* (Gilchrist & Stokes) Carmichael)
   ascigerous state: *Ajellomyces*
   2 species (Fig. 131b)
   Because the type species of *Blastomyces* was uncertain, the genus name *Zymonema* is used by Dodge (1935). A related genus is *Paracoccidioides* Almeida.
   References: Conant, 1954; Carmichael, 1962; Emmons et al., 1963.

28. **Microsporon** Gruby - C. r. Séanc. hebd. Acad. Sci. Paris *17*: 301 (1834)
   M. audouinii Gruby
   synonyms vide Dodge, 1935
   ascigerous states: *Nannizzia, Arthroderma*
   about 12 species (Fig. 131d)
   References: Dodge, 1935; Conant, 1954; Vanbreuseghem, 1952; Georg et al., 1962; Emmons et al., 1963.

29. **Keratinomyces** Vanbreuseghem - Bull. Acad. R. Med. Belg. *38*: 1068 (1952)
   K. ajelloi Vanbreuseghem
   ascigerous state: *Arthroderma*
   1 species, usually soil borne.

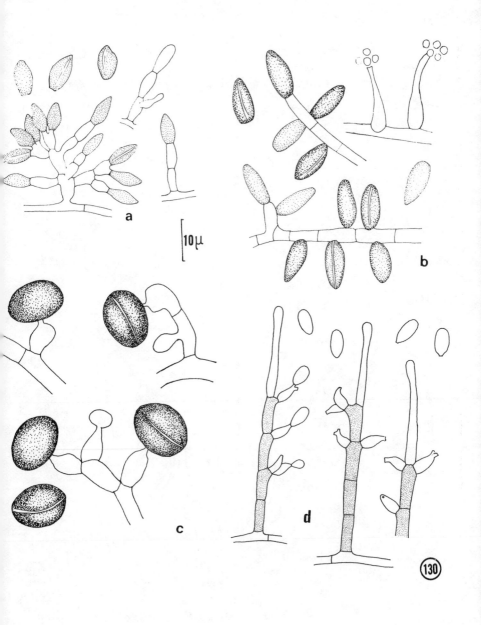

Fig. 130. a, *Wardomyces anomalus*; b, *Mammaria echinobotryoides*; c, *Nigrospora oryzae*; d, *Zygosporium masonii* Hughes (a from Hennebert, 1962; b, c, d orig.).

257

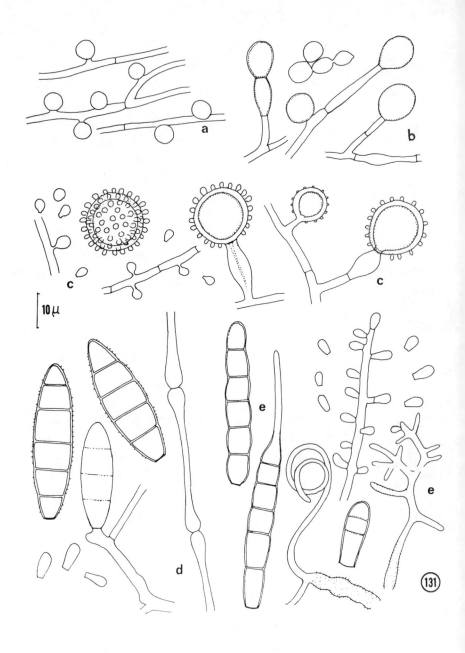

Fig. 131. a, *Trichosporiella hyalina*; b, *Blastomyces dermatitidis*; c, *Histoplasma capsulatum*; d, *Microsporum gypseum* (Bodin) Guiart & Grigorakis; e, *Trichophyton mentagrophytes* (Robin) Blanchard (orig., a from W. Gams. b-e from G. A. de Vries).

0. Trichophyton Malmsten - Arch. Anat. Phys. Wiss. Med. p. 1 (1848)
    T. tonsurans Malmsten (= *T. crateriforme* Sabouraud)
    synonyms vide Dodge, 1935
    ascigerous state: *Arthroderma*
    about 40 species, some of them are difficult to distinguish. The genus needs a
    taxonomic revision (Fig. 131e).
    References: Emmons et al., 1963; Frey, 1965; Dodge, 1935; Conant, 1954;
    Rebell et al., 1964.

1. Epidermophyton Sabour. - Les maladies du Cuir Chevelu *3*: 420 (1910)
    E. floccosum (Harz) Langeron
    1 species with pyriform macroconidia, parasitic on man.
    References: see *Trichophyton*

2. Stigmina Sacc. - Michelia *2*: 22 (1880)
    S. platani (Fuckel) Sacc.
    = *Thyrostroma* Höhnel - Sber. K. Akad. Wiss. Wien, math.-nat. Kl. *120*: 472 (1911)
        T. compactum (Sacc.) Höhnel = *S. compacta* (Sacc.) M. B. Ellis
    about 40 species, mostly known only as herbarium specimens
    References: Ellis, 1959, 1961; Hughes, 1952.

3. Annellophorella Subram. - Proc. Indian Acad. Sci., B, *55*: 6 (1962)
    A. faureae (P. Henn.) M.B. Ellis
    1 species.

4. Brachysporiella Batista - Bolm Secr. Agr. Ind. Com. Est. Pernambuco *19*: 108
    (1952)
    B. gayana Batista (=*Edmundmasonia pulchra* Subram.)
    3 species
    References: Ellis, 1959, 1971.

5. Ceratophorum Sacc. - Michelia *2*: 22 (1880)
    C. helicosporum (Sacc.) Sacc.
    2 species
    References: Ellis, 1958; Hughes, 1951.

6. Sporidesmium Link ex Fr. - Syst. mycol. *3*: 492 (1832)
    S. atrum Link ex Fr.
    = *Podoconis* Boedijn - Bull. Jard. bot. Buitenz. 3, *13*: 133 (1933)
        P. theae (Bern.) Boedijn = *S. theae* (Bern.) Hughes
    about 70 species, mostly only known as herbarium specimens
    References: Ellis, 1958, 1961, 1965; Moore, 1958.

7. Murogenella Goos & Morris - Mycologia *57*: 776 (1965)
    M. terrophila Goos & Morris (Fig. 133a)
    1 species.

8. Trichocladium Harz - Bull. Soc. imp. Moscou *44*: 125 (1871)
    T. asperum Harz (≡ *Harziella* O. Kuntze in Rev. Gener. Plant. 1891)
    3 or 4 species (Fig. 133b)
    References: Hughes, 1952, 1959; Kendrick and Bhatt, 1966; Dixon, 1968.

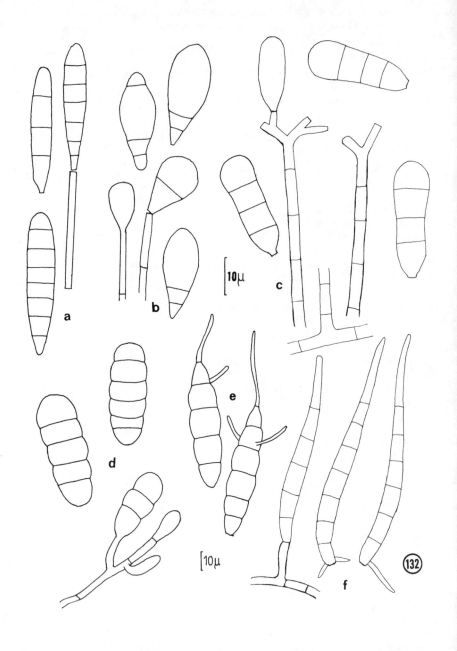

Fig. 132. a, *Dactylella minuta*; b, *Monacrosporium psychrophilum* (Drechsler) R. Cooke &
Dickinson; c, *Candelabrella cylindrospora* R. C. Cooke; d, *Mastigosporium deschampsiae*
Jørstad; e, *Mastigosporium album*; f, *Centrospora acerina* (a orig. from W. Gams, b from
Domsch and Gams, 1970, c-f orig.).

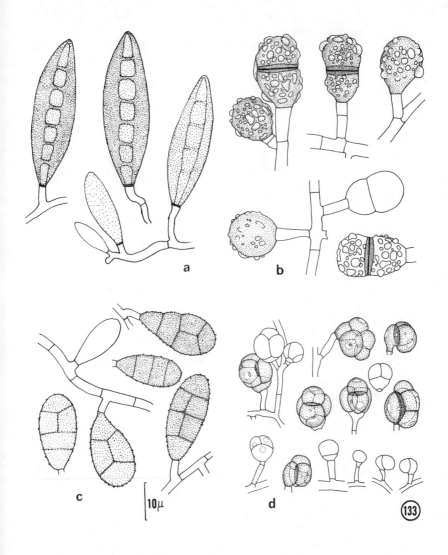

Fig. 133. a, *Murogenella terrophila*; b, *Trichocladium asperum*; c, *Pithomyces chartarum*; d, *Monodictys levis* (Wiltshire) Hughes (a, c orig., b, d from Domsch and Gams, 1970).

39. **Pithomyces** Berk. & Br. - J. Linn. Soc. Lond. *14*: 100 (1873)
     P. flavus Berk. & Br.
   = *Biconiosporium* Batista & Bezerra - Publ. Inst. Micol. Univ. Recife *417*: 4 (1964)
     B. baccharidis Batista & Bezerra = *P. quadratus* (Atk.) M. B. Ellis
   = *Neomichelia* Penz. & Sacc. - Malpighia *15*: 246 (1902)
     N. melaxantha Penz. & Sacc. = *P. flavus* Berk. & Br.
   = *Scheleobrachea* Hughes - Can. J. Bot. *36*: 802 (1958)
     Sch. echinulata (Speg.) Hughes = *P. chartarum* (Berk. & Curt.) M. B. Ellis

about 10 species (Fig. 133c)
References: Ellis, 1960; Dingley, 1962, 1968.

40. **Monodictys** Hughes - Can. J. Bot. *36*: 785 (1958)
    M. putredinis (Wallr.) Hughes
    5 species (Fig. 133d)
    References: Ellis, 1971.

41. **Kostermansinda** Rifai - Reinwardtia 7: 376 (1968)
    K. magna (Boedijn) Rifai
    1 species.

42. **Listeromyces** Penz. & Sacc. - Malpighia *15*: 259 (1901)
    L. insignis Penz. & Sacc.
    1 species.

43. **Physalidium** Mosca - Allionia *11*: 78 (1965)
    P. elegans Mosca
    1 species.

Moniliales Key VI (Watermoulds, Helicosporae, Staurosporae, etc.)

List of genera

(General references: Nilsson, 1964; Moore, 1955; Linder, 1929)

1. **Hobsonia** Berk. - Ann. Bot. *5*: 509 (1891)
   H. gigaspora Berk.
   2 species
   References: Moore, 1957; Linder, 1929.

2. **Helicoceras** Linder - Ann. Mo. bot. Gard. *18*: 3 (1931)
   H. celtidis (Biv.-Bern.) Linder
   4 species
   References: Moore, 1955; Subramanian, 1956.

3. **Helicodendron** Peyr. - Nuovo G. bot. ital. *25*: 460 (1918)
   H. paradoxum Peyr.
   8 species (Fig. 134a)
   References: van Beverwijk, 1953; Barron, 1961; Glenn-Bott, 1951.

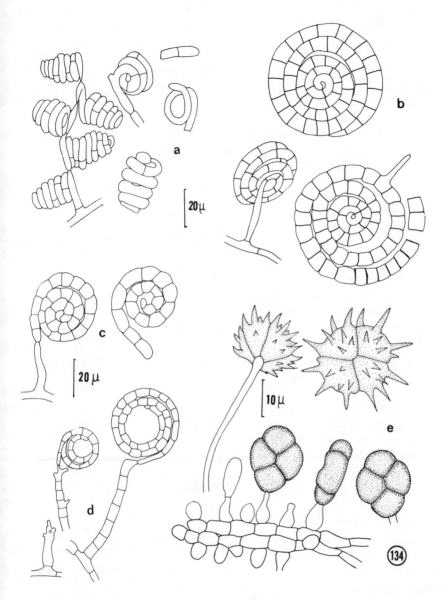

Fig. 134. a, *Helicodendron westerdijkiae* van Beverwijk; b, *Helicoon pluriseptatum* van Beverwijk; c, *Helicomyces ambiguus* (Morgan) Linder; d, *Helicosporium panacheum* Moore; e, *Spegazzinia tessarthra* (a, b orig. from W. Gams, c, d from Tubaki, 1958; e orig.).

4. Helicoon Morgan - J. Cincinnati Soc. nat. Hist. *15*: 49 (1892)
    H. sessile Morgan
10 species (Fig. 134b)
References: Linder, 1929; van Beverwijk, 1953, 1954; Litvinov, 1967.

5. **Helicoma** Corda - Icon. Fung. *1*: 15 (1837)
   H. muelleri Corda
   about 35 species
   ascigerous state: *Lasiosphaeria*
   References: Linder, 1929; Moore, 1955; Kendrick, 1958.

6. **Helicomyces** Link ex Link - Linn. Spec. Pl. *1*: 131 (1824)
   H. roseus Link ex Link
   8 species (Fig. 134c)
   References: Linder, 1929; Moore, 1954, 1955, 1957.

7. **Helicosporium** Nees ex Fr. - Syst. mycol. *3*: 353 (1832)
   H. vegetum Nees ex Fr.
   17 species (Fig. 134d)
   ascigerous state: *Ophionectria* Sacc.
   References: Moore, 1954, 1955.

8. **Everhartia** Sacc. & Ellis - Michelia *2*: 580 (1882)
   E. hymenuloides Sacc. & Ellis
   1 species
   References: Moore, 1954.

9. **Spirosphaera** van Beverwijk - Trans. Br. mycol. Soc. *36*: 120 (1953)
   S. floriformis van Beverwijk
   3 species
   References: Hennebert, 1968.

10. **Clathrosphaerina** van Beverwijk - Trans.Br. mycol. Soc. *34*: 289 (1951)
    C. zalewskii van Beverwijk
    1 species
    References: Tubaki, 1958.

11. **Candelabrum** van Beverwijk - Antonie van Leeuwenhoek *17*: 11 (1951)
    C. spinulosum van Beverwijk
    second species: *C. japonense* Tubaki (1958).

12. **Spegazzinia** Sacc. - Michelia *2*: 37 (1880)
    S. tessarthra (Berk. & Curt.) Sacc. (Fig. 134e)
    4 or 5 species
    References: Hughes, 1953; Damon, 1953.

13. **Actinospora** Ingold - Trans. Br. mycol. Soc. *35*: 66 (1952)
    A. megalospora Ingold
    1 species.

14. **Anguillospora** Ingold - Trans. Br. mycol. Soc. *25*: 402 (1942)
    A. longissima (de Wild.) Ingold
    ascigerous state: *Mollisia* (Fr.) Karst.
    4 species
    References: Ingold, 1958; Webster, 1961; Nilsson, 1964; Petersen, 1962.

15. **Articulospora** Ingold - Trans. Br. mycol. Soc. *25*: 376 (1942)
    A. tetracladia Ingold
    2 species

References: Petersen, 1962; Nilsson, 1963.

6. **Clavariopsis** de Wild. - Ann. Soc. Belg. Micr. *19*: 197 (1895)
   C. aquatica de Wild.
   3 species
   References: Ingold, 1942; Tubaki, 1957; Anastasiou, 1962.

7. **Lunulospora** Ingold - Trans. Br. mycol. Soc. *25*: 408 (1942)
   L. curvula Ingold.

8. **Tetracladium** de Wild. - Ann. Soc. Belg. Micr. *17*: 35 (1899)
   T. marchalianum de Wild.
   3 species
   References: Nilsson, 1964; Ingold, 1942.

9. **Tricladium** Ingold - Trans. Br. mycol. Soc. *25*: 388 (1942)
   T. splendens Ingold
   6 species, and compare *Alatospora*
   References: Nilsson, 1964.

0. **Tetrachaetum** Ingold - Trans. Br. mycol. Soc. *25*: 380 (1942)
   T. elegans Ingold
   References: Nilsson, 1964.

1. **Varicosporium** Kegel - Ber. dt. bot. Ges. *24*: 213 (1906)
   V. elodeae Kegel
   2 species
   References: Barron, 1968.

2. **Tridentaria** Preuss - Linnaea *25*: 74 (1852)
   T. alba Preuss
   2 or 3 species, partly parasitic on nematodes
   References: Drechsler, 1937, 1940.

3. **Diplocladiella** Arnaud - Bull. Soc. mycol. Fr. *69*: 296 (1953)
   D. scalaroides Arnaud
   1 species
   References: Tubaki, 1958.

4. **Triposporium** Corda - Icon. Fung. *1*: 16 (1837)
   T. elegans Corda
   2 or 3 species
   References: Hughes, 1951; Ellis et al., 1951.

5. **Tripospermum** Speg. - Physis *4*: 295 (1918)
   T. acerinum (Syd.) Speg.
   References: Hughes, 1951; Ingold and Cox, 1957.

6. **Speira** Corda - Icon. Fung. *1*: 9 (1837)
   S. toruloides Corda
   References: Hughes, 1952.
   Damon (1952) regards *Speira* as a synonym of *Dictyosporium*.

7. **Dictyosporium** Corda - Weitenweber Beitr. Natur u. Heilwiss. Prag *1*: 87 (1836)
   D. elegans Corda
   References: Damon, 1952; Subramanian, 1953.

267

Moniliales Key VII (sterile mycelia)

1. Cellular bodies (bulbils) present, usually brownish 2
1. Cellular bodies absent, true sclerotia may be present 5
2. Cellular bodies composed of few large central and numerous smaller surrounding cells . . . . *Papulaspora* (1)
2. Cellular bodies composed of uniform cells or hyphal elements 3
3. Cellular bodies with a multihyphal base, rounded above . . . . *Minimedusa* (2)
3. Cellular bodies spherical or irregular, usually formed on a hyphal branch . . 4
4. Clamp connexions absent . . . . *Myriococcum* (3)
4. Clamp connexions present . . . . *Burgoa* (4)
5. Clamp connexions present . . . . sterile Basidiomycetes
5. Clamp connexions absent 6
6. Hyphae wide, constricted at septa, with dolipores, colonies spreading quickly, sclerotia absent or present . . . . *Rhizoctonia* (6)
6. Hyphae usually without constrictions, sclerotia present . . . . . *Sclerotium* (5)

List of genera

1. **Papulaspora** Preuss - Linnaea *24*: 113 (1851)
   P. sepedonioides Preuss
   about 15 species (Fig. 135)
   References: Hotson, J.W., 1917; Hotson, H.H., 1942; van Beverwijk, 1954; Nicot, 1948.

2. **Minimedusa** Weresub & Le Clair - Can. J. Bot. *49*: 2210 (1971)
   M. polyspora (Hotson) Weresub & Le Clair
   1 or 2 species.

3. **Myriococcum** Fr. - Syst. mycol. *2*: 304 (1823)
   M. praecox Fr. (= *Papulaspora byssina* Hotson)
   second species: *Myriococcum thermophilum* (Fergus) van der Aa.

4. **Burgoa** Goidanich - Boll. Sta. Patol. Veg. Roma, n.s. *17*: 305 (1938)
   B. verzuoliana Goidanich
   3 or 4 species, imperfect states of *Sistotrema* and *Hyphodontia*
   References: Weresub and Le Clair, 1971.

5. **Sclerotium** Tode ex Fr. - Syst. mycol. *2*: 246 (1823)
   S. complanatum Tode ex Fr.
   about 15 species
   Important species are *S. cepivorum* Berk., *S. tuliparum* Kleb. and *S. wakkeri* Boerema & Posthumus on Liliaceae. *S. rolfsii* Sacc. is the sclerotial state of *Cor-*

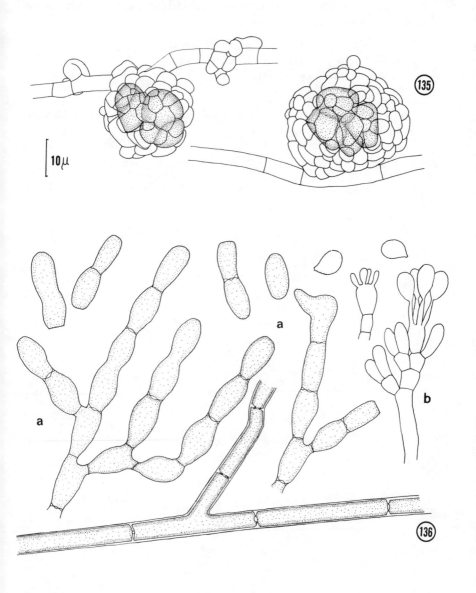

Fig. 135. *Papulaspora immersa* Hotson (from Domsch and Gams, 1970).
Fig. 136. *Rhizoctonia solani* Kühn, a, hyphae and hyphal cells; b, basidia and basidiospores of *Thanatephorus cucumeris* (Frank) Donk, the basidial state (orig.).

ticium rolfsii Curzi and the hyphae are provided with clamp connexions. In pure culture species of genera such as *Sclerotinia, Macrophomina, Typhula* or *Colletotrichum* often develop only sclerotia.
References: Purdy, 1955.

6. **Rhizoctonia** DC. ex Fr. - Syst. mycol. *2*: 265 (1823)
   R. crocorum (Pers.) DC. ex Fr.
   = *Moniliopsis* Ruhland - Arb. K. biol. Anst. *6*: 76 (1908)
       M. aderholdii Ruhland = *R. solani* Kühn
Basidial states: *Thanatephorus* Donk, *Botryobasidium* Donk, *Helicobasidium* Pat., *Pellicularia* Cooke (= *Koleroga* Donk).
A common parasitic species is *Rhizoctonia solani* Kühn (= *Thanatephorus cucumeris* (Frank) Donk (Fig. 136). The basidial state of *R. crocorum* is *Helicobasidium purpureum* (Tul.) Pat.
References: Saksena and Vaartaya, 1960, 1961; Donk, 1958; Warcup and Talbot, 1966, 1967.

REFERENCES

AA, H. A. van der (1968) - Petrakia irregularis, a new fungus species. Acta bot. neerl. *17*: 221-225.

AA, H.A. van der (1973) - Studies in Phyllosticta I. Stud. Mycol. *5*, 110 pp.

AA, H.A. van der & H.A. van KESTEREN (1971) - The identity of Phyllosticta destructiva and similar Phoma-like fungi described from Malvaceae and Lycium halimifolium. Acta bot. neerl. *20*: 552-563.

AEBI, B. (1972) - Untersuchungen über Discomyceten aus der Gruppe Tapesia-Trichobelonium. Nova Hedwigia *23*: 49-112.

AHMED, S.I. & R.F. CAIN (1972) - Revision of the genera Sporormia and Sporormiella. Can. J. Bot. *50*: 419-477.

AINSWORTH, G. C. (1961) - Ainsworth & Bisby's Dictionary of Fungi. C.M.I., Kew, 547 pp.

AINSWORTH, G. C. & K. SAMPSON (1950) - The British smut fungi. C.M.I., Kew, 137 pp.

AJELLO, L. (1959) - A new Microsporum and its occurrence in soil and on animals. Mycologia *51*: 69-76.

AJELLO, L. (1968) - A taxonomic review of the dermatophytes and related species. Sabouraudia 6: 147-159.

ALLESCHER, A. (1901 and 1903) - Fungi imperfecti in Rabenhorst Krypt.Fl., Pilze VI & VII, Leipzig, 1016 & 1072 pp.

AMES, L. M. (1963) - A monograph of the Chaetomiaceae. U.S. Army Res. Dev. Ser. 2, 125 pp.

AMMON, U. U. (1963) - Über einige Arten aus den Gattungen Pyrenophora und Cochliobolus mit Helminthosporium als Nebenfruchtform. Phytopath. Z. *47*: 244-300.

ANASTASIOU, C. J. (1961) - Fungi from salt lakes. I. A new species of Clavariopsis. Mycologia *53*: 11-16.

ANASTASIOU, C. J. (1963) - Fungi from salt lakes. Nova Hedwigia 6: 243-276.

APINIS, A. E. (1964) - Revision of British Gymnoascaceae. Mycol. Pap. *96*, 56 pp.

APINIS, A. E. (1967) - Dactylomyces and Thermoascus. Trans. Br. mycol. Soc. *50*: 573-582.

APINIS, A.E. (1968) - Relationships of certain keratinophilic Plectascales. Mycopath. Mycol. appl. *35*: 97-104.

APINIS, A. E. & C. G. C. CHESTERS (1964) - Ascomycetes of some salt marshes and sand dunes. Trans. Br. mycol. Soc. *47*: 419-435.

APINIS, A. E. & H. O. W. EGGINS (1966) - Thermomyces ibadanensis from oil palm kernel stacks in Nigeria. Trans. Br. mycol. Soc. *49*: 629-632.

ARCHER, W. A. (1926) - Morphological characters of some Sphaeropsidales in culture. Ann. mycol. *24*: 1-84.

ARNOLD, G.R.W. (1969) - Bestimmungsschlüssel für die wichtigsten und häufigsten mykophilen Ascomyceten und Hyphomyceten. Z. Pilzk. *35*: 41-45.

ARNOLD, G.R.W. (1971) - Zur Systematik der Hypomycetaceae. Nova Hedwigia *21*: 529-535.

ARX, J. A. von (1949) - Beiträge zur Kenntnis der Gattung Mycosphaerella. Sydowia *3*: 28-100.

ARX, J. A. von (1952) - Über einige Ascomycetengattungen mit ungleich zweizelligen Sporen. Ber. schweiz. bot. Ges. *62*: 340-362.

ARX, J. A. von (1952) - Studies on Venturia and related genera. Tijdschr. Pl.Ziekt. *58*: 260-266.

ARX, J. A. von (1955) - Ein neuer Ascomycet aus Afrika. Antonie van Leeuwenhoek *21*: 161-165.

ARX, J. A. von (1957) - Schurft of Pyracantha; Über Fusicladium saliciperdum. Tijdschr. Pl. Ziekt. *63*: 198-199, 232-236.

ARX, J. A. von (1957) - Die Arten der Gattung Colletotrichum. Phytopath. Z. *29*: 413-468.

ARX. J. A. von (1957) - Revision der zu Gloeosporium gestellten Pilze. Verh. K. ned. Akad. Wet., afd. Natuurk., 2de reeks, *51* (3), 153 pp.

ARX, J. A. von (1961) - Über Cylindrosporium padi. Phytopath. Z. *42*: 161-166.

ARX, J. A. von (1963) - Revision der zu Gloeosporium gestellten Pilze. Nachträge und Berichtigungen. Proc. K. ned. Akad. Wet. C, *66*: 172-182.

ARX, J. A. von (1963) - Die Gattungen der Myriangiales. Persoonia *2*: 421-475.

ARX, J. A. von (1967) - Pilzkunde. Ein kurzer Abriss der Mykologie unter besonderer Berücksichtigung der Pilze in Reinkultur. J. Cramer, Lehre, 356 pp.

ARX, J. A. von (1970) - A Revision of the Fungi classified as Gloeosporium. J. Cramer, Lehre, 203 pp.

ARX, J.A. von (1971) - Über die Typusart, zwei neue und einige weitere Arten der Gattung Sporotrichum. Persoonia *6*: 179-184.

ARX, J.A. von (1971) - Testudinaceae, a new family of Ascomycetes. Persoonia *6*: 365-369.

ARX, J.A. von (1971) - On Arachniotus and related genera of the Gymnoascaceae. Persoonia *6*: 371-380.

ARX, J.A. von (1972) - On Endomyces, Endomycopsis and related yeast-like fungi. Antonie van Leeuwenhoek *38*: 289-309.

ARX, J.A. von (1973) - Further observations on Sporotrichum and some similar fungi. Persoonia *7*: 127-130.

ARX, J.A. von (1973) - Ostiolate and nonostiolate Pyrenomycetes. Proc. K. ned. Akad. Wet., Sect. C, *76*: 289-296.

ARX, J.A. von (1973) - The genera Petriellidium and Pithoascus (Microascaceae). Persoonia *7*: 367-375.

ARX, J. A. von & W. GAMS (1966) - Über Pleurage verruculosa und die zugehörige Cladorrhinum-Konidienform. Nova Hedwigia *13*: 199-208.

ARX, J. A. von & G. L. HENNEBERT (1968) - Triangularia mangenotii nov. sp. Bull. Soc. mycol. Fr. *84*: 423-426.

ARX, J. A. von & E. MÜLLER (1954) - Die Gattungen der amerosporen Pyrenomyceten. Beitr. Krypt.Fl. Schweiz, *11*, 1, 434 pp.

ARX, J. A. von & P. K. STORM (1967) - Über einige aus dem Erdboden isolierte, zu Sporormia, Preussia und Westerdijkella gehörende Ascomyceten. Persoonia *4*: 407-415.

AUSTWICK, P. K. C. (1954) - Mastigosporium deschampsiae in Great Britain. Trans. Br. mycol. Soc. *37*: 161-165.

BACKUS, M. P. & P. A. ORPURT (1961) - A new Emericellopsis from Wisconsin, with notes on other species. Mycologia *53*: 64-83.

BAINIER, G. (1907) - Mycothèque de l'École de Pharmacie. XIV. Scopulariopsis (Penicillium pro parte) genre nouveau de Mucédinées. Bull. Soc. mycol. Fr. *23*: 98-105.

BAIJAL, U. & B.S. MEHROTRA (1965) - Species of Mucor from India - II. Sydowia *19*: 204-212.

BAKSHI, B. K. (1952) - Oedocephalum lineatum is a conidial stage of Fomes annosus. Trans. Br. mycol. Soc. *35*: 195.

BALDACCI, E., R. CIFERRI & E. VACCARI (1938) - Revisione sistematica del genere Malbranchea. Atti Ist. bot. Lab. critt. Pavia, Ser. 4, *11*: 75-103 (1939).

BARNETT, H. L. (1958) - A new Calcarisporium parasitic on other fungi. Mycologia *50*: 497-500.

BARR, D.J.S. (1969) - Studies in Rhizophydium and Phlyctochytrium. Can. J. Bot. *47*: 991-997.

BARR, D.J.S. (1970) - Hyphochytrium catenoides: a morphological and physiological study of

North American isolates. Mycologia *62*: 492-503.

BARR, D.J.S. (1973) - Six Rhizophydium species in culture. Can. J. Bot. *51*: 967-975.

BARR, M. E. (1958) - Life history studies on Mycosphaerella tassiana and M. typhae. Mycologia *50*: 501-513.

BARR, M. E. (1964) - The genus Pseudomassaria in North America. Mycologia *56*: 841-862.

BARR, M. E. (1968) - The Venturiaceae in North America. Can. J. Bot. *46*: 799-864.

BARRON, G. L. (1961) - Studies on species of Oidiodendron, Helicodendron, and Stachybotrys from soil. Can. J. Bot. *39*: 1563-1571.

BARRON, G. L. (1962) - New species and new records of Oidiodendron. Can. J. Bot. *40*: 589-607.

BARRON, G. L. (1964) - A note on the relationship between Stachybotrys and Hyalostachybotrys. Mycologia *56*: 313-316.

BARRON, G. L. (1968) - The genera of Hyphomycetes from soil. Williams & Wilkins, Baltimore, 364 pp.

BARRON, G. L. & G. C. BHATT (1967) - A new species of Gonytrichum from soil. Mycopath. Mycol. appl. *32*: 126-128.

BARRON, G. L. & L. V. BUSCH (1961) - Studies on the soil hyphomycete Scolecobasidium. Can. J. Bot. *40*: 77-84.

BARRON, G. L. , R. F. CAIN & J. C. GILMAN (1961) - A revision of the genus Petriella. Can. J. Bot. *39*: 837-845.

BARRON, G. L., R. F. CAIN & J. C. GILMAN (1961) - The genus Microascus. Can. J. Bot. *39*: 1609-1631.

BARRON, G. L. & A. H. S. ONIONS (1966) - Verticillium chlamydosporium and its relationships to Diheterospora, Stemphyliopsis and Paecilomyces. Can. J. Bot. *44*: 861-869.

BATEMAN, D. F. (1963) - Influence of host and non-host plants upon populations of Thielaviopsis basicola in soil. Phytopathology *53*: 1174-1177.

BATISTA, A.C. & R. CIFERRI (1963) - Capnodiales. Saccardoa 2, 298 pp.

BATRA, L. R. (1959) - A comparative morphological and physiological study of the species of Dipodascus. Mycologia *51*: 329-355.

BATRA, L. R. (1967) - Ambrosia Fungi: A taxonomic revision and nutritional studies of some species. Mycologia *59*: 976-1017.

BATRA, L. R. & H. FRANCKE—GROSMANN (1964) - Two new Ambrosia Fungi - Ascoidea asiatica and A. africana. Mycologia *56*: 632-636.

BENHAM, R. W. & J. L. MIRANDA (1953) - The genus Beauveria, morphological and taxonomical studies of several species and of two strains isolated from wharf-piling borers. Mycologia *45*: 727-746.

BENJAMIN, C. R. (1955) - Ascocarps of Aspergillus and Penicillium. Mycologia *47*: 669-687.

BENJAMIN, C. R. & C. W. HESSELTINE (1957) - The genus Actinomucor. Mycologia *49*: 240-249.

BENJAMIN, C. R. & C. W. HESSELTINE (1959) - Studies on the genus Phycomyces. Mycologia *51*: 751-771.

BENJAMIN, R. K. (1955) - An addition to the genus Magnusia. Aliso *3*: 199-201.

BENJAMIN, R. K. (1956) - A new genus of the Gymnoascaceae with a review of the other genera. Aliso *3*: 301-328.

BENJAMIN, R. K. (1958) - Sexuality in the Kickxellaceae. Aliso *4*: 149-169.

BENJAMIN, R. K. (1959) - The merosporangiferous Mucorales. Aliso *4*: 321-453.

BERTHET, P. (1964) - Formes conidiennes de divers Discomycètes. Bull. trimest. Soc. mycol. Fr. *80*: 125-149.

BESTAGNO BIGA, M. L., R. CIFERRI & G. BESTAGNO (1958) - Ordinamento artificiale delle specie del genere Coniothyrium. Sydowia *12*: 258-320.

BEVERWIJK, A. L. van (1953) - Helicosporous Hyphomycetes I. Trans. Br. mycol. Soc. *36*: 111-124.

BEVERWIJK, A. L. van (1954) - Three new fungi. Antonie van Leeuwenhoek *20*: 1-16.

BEYMA thoe KINGMA, F. H. van (1942) - Beschreibung einiger neuer Pilzarten aus dem Centraalbureau voor Schimmelcultures, Baarn (Nederland). Antonie van Leeuwenhoek *8*: 105-122.

BEYMA thoe KINGMA, F. H. van (1943) - Beschreibung der im Centraalbureau voor Schimmelcultures vorhandenen Arten der Gattungen Phialophora und Margarinomyces nebst Schlüssel zu ihrer Bestimmung. Antonie van Leeuwenhoek *9*: 51-76.

BEZERRA, J. L. (1963) - Studies on Pseudonectria rousseliana. Acta bot. neerl. *12*: 58-63.

BHATT, G. C. & W. B. KENDRICK (1968) - The generic concepts of Diplorhinotrichum and Dactylaria, and a new species of Dactylaria from soil. Can. J. Bot. *46*: 1253-1257.

BJÖRLING, K. (1936) - Über die Gattungen Mortierella und Haplosporangium. Bot. Notiser p. 116-126.

BLOCHWITZ, A. (1914) - Botryotrichum piluliferum. Ann. mycol. *12*: 315-334.

BLUMER, S. (1958) - Beiträge zur Kenntnis von "Cylindrosporium padi". Phytopath. Z. *33*: 263-290.

BOEDIJN, K. B. (1927) - Über Rhopalomyces elegans. Ann. mycol. *25*: 161-166.

BOEDIJN, K. B. (1958) - Notes on the Mucorales of Indonesia. Sydowia *12*: 321-362.

BOEDIJN, K. B. (1962) - The Sordariaceae of Indonesia. Persoonia *2*: 305-320.

BOEDIJN, K. B. & J. REITSMA (1950) - Notes on the genus Cylindrocladium (Fungi: Mucedinaceae). Reinwardtia *1*: 51-60.

BOEREMA, G. H. (1964) - Phoma herbarum, the type-species of the form-genus Phoma. Persoonia *3*: 9-16.

BOEREMA, G. H. (1970) - Additional notes on Phoma herbarum. Persoonia *6*: 15-48.

BOEREMA, G. H. & M. M. J. DORENBOSCH (1965) - Phoma-achtige schimmels in associatie met appelbladvlekken. Versl. Pl.Ziekt. Dienst Wageningen *142*: 138-151.

BOEREMA, G. H. & M. M. J. DORENBOSCH (1970) - On Phoma macrostomum Mont., an ubiquitous species on woody plants. Persoonia *6*: 49-58.

BOEREMA, G. H. & M. M. J. DORENBOSCH (1973) - The Phoma and Ascochyta species described by Wollenweber and Hochapfel in their study on fruit-rotting. Stud. Mycol. *3*, 50 pp.

BOEREMA, G. H., M. M. J. DORENBOSCH & H. A. van KESTEREN (1965) - Remarks on species of Phoma referred to Peyronellaea. Persoonia *4*: 47-68.

BOEREMA, G. H., M. M. J. DORENBOSCH & H. A. van KESTEREN (1968) - Remarks on species of Phoma referred to Peyronellaea - II. Persoonia *5*: 201-205.

BOEREMA, G. H. & L. H. HÖWELER (1967) - Phoma exigua and its varieties. Persoonia *5*: 15-28.

BOIDIN, J., M. C. PIGNAL & M. BESSON (1965) - Le genre Pichia sensu lato. Bull. Soc. mycol. Fr. *81*: 567.

BOLLARD, E. G. (1950) - Studies on the genus Mastigosporium. Trans. Br. mycol. Soc. *33*: 250-275.

BONNER, J. T. (1967) - The cellular slime molds. 2nd edit., Princetown, 205 pp.

BOOTH, C. (1957) - Studies of Pyrenomycetes I. Four species of Chaetosphaeria, two with Catenularia conidia. Mycol. Pap. *68*, 27 pp.

BOOTH, C. (1958) - The genera Chaetosphaeria and Thaxteria in Britain. Naturalist, Hull. p. 83-90.

274

BOOTH, C. (1959) - Studies of Pyrenomycetes IV. Nectria. Mycol. Pap. *73*, 115 pp.

BOOTH, C. (1960) - Studies of Pyrenomycetes V. Nomenclature of some Fusaria in relation to their nectrioid perithecial states. Mycol. Pap. *74*, 16 pp.

BOOTH, C. (1961) - Studies of Pyrenomycetes VI. Thielavia with notes on some allied genera. Mycol. Pap. *83*, 15 pp.

BOOTH, C. (1966) - The genus Cylindrocarpon. Mycol. Pap. *104*, 56 pp.

BOOTH, C. (1971) - The genus Fusarium. CMI, Kew, 237 pp.

BOOTH, C. & J. S. MURRAY (1960) - Calonectria hederae and its Cylindrocladium conidial state. Trans. Br. mycol. Soc. *43*: 69-72.

BOOTH, C. & W. A. SHIPTON (1966) - Thielavia pilosa sp. nov., with a key to species of Thielavia. Trans. Br. mycol. Soc. *49*: 665-667.

BOSE, S. K. (1961) - Studies on Massarina and related genera. Phytopath. Z. *41*: 151-213.

BOWEN, P. R. (1930) - A maple leaf disease caused by Cristulariella depraedans. Bull. Conn. agric. Exp. Stn *316*: 625-647.

BREWER, J. G. & G. H. BOEREMA (1965) - Electron microscope observations on the development of pycnidiospores in Phoma and Ascochyta. Proc. K. ned. Akad. Wet. C. *68*: 86-97.

BROWN, A. H. S. & G. SMITH (1957) - The genus Paecilomyces and its perfect stage Byssochlamys. Trans. Br. mycol. Soc. *40*: 17-89.

BRUMMELEN, J. van (1967) - A World-monograph of the genera Ascobolus and Saccobolus. Diss. Leiden, 260 pp.

BUCHWALD, N. F. (1949) - Studies in the Sclerotiniaceae I. Taxonomy of the Sclerotiniaceae. Contr. Dep. Pl. Path. R. Vet. Agr. Coll. Copenhagen *32* (repr. 68-191).

BÜTIN, H. (1957) - Über zwei Arten der Gattung Cryptodiaporthe und ihre zugehörigen Nebenfruchtformen. Sydowia *11*: 27-29.

BUTIN, H. (1958) - Über die auf Salix und Populus vorkommenden Arten der Gattung Cryptodiaporthe. Phytopath. Z. *32*: 399-415.

BUTIN, H. (1963) - Über Sclerophoma pityophila als Bläuepilz an verarbeitetem Holz. Pytopath. Z. *48*: 298-305.

BUTIN, H. (1963) - Über zwei Nebenfruchtformen von Sydowia polyspora. Sydowia *17*: 114-118 (ersch. 1964).

BUTLER, E.E. & L.J. PETERSEN (1972) - Endomyces geotrichum a perfect state of Geotrichum candidum. Mycologia *64*: 365-374.

CAILLEUX, R. (1969) - Champignons stercoraux de République Centrafricaine II & III. Cahiers Maboké *7*: 5-14, 87-102.

CAILLEUX, R. (1970) - Champignons stercoraux de République Centrafricaine IV.-Tripterospora. Cahiers Maboké *8*: 5-16.

CAILLEUX, R. (1971) - Recherches sur la mycoflore coprophile centrafricaine. Les genres Sordaria, Gelasinospora, Bombardia. Bull. trimest. Soc. mycol. Fr. *87*: 461-626.

CAIN, R. F. (1934) - Studies of coprophilous Sphaeriales in Ontario. Univ. Toronto Studies no. *38*, 126 pp. Reprint J. Cramer, Lehre, 1969.

CAIN, R. F. (1950) - Studies of coprophilous Ascomycetes. I. Gelasinospora. Can. J. Res., Sect. C., *28*: 566-576.

CAIN, R. F. (1956) - Studies of coprophilous Ascomycetes IV. Tripterospora, a new cleistocarpous genus in a new family. Can. J. Bot. *34*: 699-710.

CAIN, R. F. (1957) - Studies of coprophilous Ascomycetes VI. Can. J. Bot. *35*: 255-265.

CAIN, R. F. (1961) - Studies of soil fungi III. Can. J. Bot. *39*: 1231-1239.

CAIN, R. F. (1961) - Anixiella and Diplogelasinospora, two genera with cleistothecia and pitted ascospores. Can. J. Bot. *39*: 1667-1677.

CAIN, R. F. (1961) - Studies of coprophilous Ascomycetes VII. Preussia. Can. J. Bot. *39*: 1633-1666.

CAIN, R. F. (1962) - Studies of coprophilous Ascomycetes VIII. New species of Podospora. Can. J. Bot. *40*: 447-490.

CAIN, R. F. & W. M. FARROW (1956) - Studies of coprophilous Ascomycetes III. The genus Triangularia. Can. J. Bot. *34*: 689-697.

CAIN, R. F. & J. W. GROVES (1948) - Notes on seed-borne fungi VI. Sordaria. Can. J. Res., Sect. C, *26*: 486-495.

CAIN, R.F. & J.H. MIRZA (1972) - Three new species of Arnium. Can. J. Bot. *50*: 333-336.

CAIN, R. F. & L. K. WERESUB (1957) - Studies of coprophilous Ascomycetes V. Sphaeronaemella fimicola. Can. J. Bot. *35*: 119-131.

CALDWELL, R. M. (1937) - Rhynchosporium scald of barley, rye and other grasses. J. agric. Res. *55*: 175-198.

CARMICHAEL, J. W. (1957) - Geotrichum candidum. Mycologia *49*: 820-830.

CARMICHAEL, J. W. (1962) - Chrysosporium and some other aleuriosporic Hyphomycetes. Can. J. Bot. *40*: 1137-1173.

CARRANZA, J. M. & E. S. LUTTRELL (1967) - An undescribed species of Exosporium on grain of Secale cereale. Mycologia *59*: 1097-1101.

CEJP, K. & A. A. MILKO (1964) - Genera of the Eurotiaceae with 32 ascospores. I. Westerdijkella. Česká Mykol. *18*: 82-84.

CHARLES, V. K. (1935) - A little known pecan fungus. Mycologia *27*: 74-82.

CHESTERS, C. G. C. (1935) - The life histories of three species of Cephalotheca. Trans. Br. mycol. Soc. *19*: 261-279.

CHILTON, J. E. (1954) - Volutella species on alfalfa. Mycologia *46*: 800-809.

CHUPP, C. (1953) - A monograph of the fungus genus Cercospora. New York, 667 pp.

COKER, W. C. (1923) - The Saprolegniaceae, with notes on other water molds. Univ. North Carolina Press, 201 pp.

COKER, W. C. & V. D. MATTHEWS (1937) - Blastocladiales, Monoblepharidales, Saprolegniales. N. Am. Flora *2* (1), 76 pp.

COLE, G. T. & W. B. KENDRICK (1968) - Conidium ontogeny in hyphomycetes. The imperfect state of Monascus ruber and its meristem arthrospores. Can. J. Bot. *46*: 987-992.

CONANT, N. F., D. T. SMITH, R. D. BAKER, J. L. CALLAWAY & D. S. MARTIN (1954) - Manual of clinical mycology, 2nd ed., Philadelphia, 456 pp.

COOKE, R. C. & C. H. DICKINSON (1965) - Nematode-trapping species of Dactylella and Monacrosporium. Trans. Br. mycol. Soc. *48*: 621-629.

COOKE, R. C. & B. E. S. GOODFRY (1964) - A key to the nematode-destroying fungi. Trans. Br. mycol. Soc. *47*: 61-74.

COOKE, W.B. (1968) - Studies in the genus Prototheca II. Taxonomy. J. Elisha Mitchell scient. Soc. *84*: 217-220.

COONEY, D. G. & R. EMERSON (1964) - Thermophilic Fungi. San Francisco, 188 pp.

COPELAND, H. F. (1938) - The kingdoms of organisms. Quart. Rev. Biol. *13*: 383-420.

CORBAZ, R. (1957) - Recherches sur le genre Didymella. Phytopath. Z. *28*: 375-414.

CORBETTA, G. (1965) - Rassegna delle specie del genere Curvularia. Riso *14*: 3-23.

COUCH, J. N. (1939) - A new Conidiobolus with sexual reproduction. Am. J. Bot. *26*: 119-130.

CRANE, J.L. (1972) - Illinois Fungi III. Dendrosporium lobatum and Sporidesmium taxodii. Trans. Br. mycol. Soc. *58*: 423-426.

CUNNELL, G. J. (1957) - On Neottiospora caricina. Trans. Br. mycol. Soc. 40: 433-442.

CUNNELL, G. J. (1958) - On Robillarda phragmitis sp. nov. Trans. Br. mycol. Soc. 41: 405-412.

CUTTER, V. M. (1941) - Observations on certain species of Aphanomyces. Mycologia 33: 220-240.

CUTTER, V. M. (1946) - The genus Cunninghamella. Farlowia 2: 321-343.

DAMON, S. C. (1952) - Two noteworthy species of Sepedonium. Mycologia 44: 86-96.

DAMON, S. C. (1952) - Type studies in Dictyosporium, Speira and Cattanea. Lloydia 15: 110-124.

DAMON, S. C. (1953) - Notes on the hyphomycetous genera Spegazzinia and Isthmospora. Bull. Torrey bot. Club 80: 155-165.

DAVIDSON, R. W. (1958) - Additional species of Ophiostomataceae from Colorado. Mycologia 50: 661-670.

DAVIDSON, R. W., T. E. HINDS & E. R. TOOLE (1964) - Two new species of Ceratocystis from hardwoods. Mycologia 56: 793-798.

DAWSON, C. O. (1963) - Two new species of Arthroderma isolated from soil rabbit burrows. Sabouraudia 2: 185-191.

DAWSON, C. O. & J. C. GENTLES (1961) - The perfect states of Keratinomyces ajelloi, Trichophyton terrestre and Microsporum nanum. Sabouraudia 1: 49-57.

DÉFAGO, G. (1935) - De quelques Valsées parasites des arbres à noyau dépérissants. Beitr. Krypt.Fl. Schweiz 8, 3, 109 pp.

DEIGHTON, F. C. (1959) - Studies on Cercospora and allied genera. Mycol. Pap. 71, 23 pp.

DEIGHTON, F. C. (1960) - African Fungi I. Mycol. Pap. 78, 43 pp.

DEIGHTON, F.C. (1967) - Studies on Cercospora and allied genera II. Mycol. Pap. 112, 88 pp.

DEIGHTON, F. C. (1968) - Spermospora. Trans. Br. mycol. Soc. 51: 41-49.

DEIGHTON, F.C. (1971) - Studies on Cercospora and allied genera III. Centrospora. Mycol. Pap. 124, 13 pp.

DENNIS, R. W. G. (1956) - A revision of the British Helotiaceae in the herbarium of the Royal Botanic Gardens, Kew, with notes on related European species. Mycol. Pap. 62, 216 pp.

DENNIS, R. W. G. (1968) - British Ascomycetes. J. Cramer, Lehre, 455 pp.

DERX, H. G. (1930) - Études sur les Sporobolomycètes. Ann. mycol. 28: 1-23.

DICKINSON, C. H. (1964) - The genus Wardomyces. Trans. Br. mycol. Soc. 47: 321-325.

DICKINSON, C. H. (1966) - Wardomyces pulvinata comb. nov. Trans. Br. mycol. Soc. 49: 521-522.

DICKINSON, C. H. (1968) - Gliomastix. Mycol. Pap. 115, 24 pp.

DIEDICKE, H. (1915) - Pilze VII. Sphaeropsideae, Melanconiae. Krypt.Fl. Mark Brandenbg. 9, 962 pp.

DINGLEY, J. M. (1951) - The Hypocreales of New Zealand II: The genus Nectria. Trans. R. Soc. N. Z. 79: 177-202.

DINGLEY, J. M. (1952) - The genus Hypocrea. Trans. R. Soc. N. Z. 79: 323-337.

DINGLEY, J. M. (1957) - Life history studies of New Zealand species of Nectria. Tans. R. Soc. N. Z. 84: 467-477.

DINGLEY, J. M. (1957) - Life history studies in the genus Hypocrea. Trans. R. Soc. N. Z. 84: 689-693.

DINGLEY, J. M. (1962) - Pithomyces chartarum, its occurrence, morphology and taxonomy. N. Z. Jl agric. Res. 5: 49-61.

DIXON, M. (1968) - Trichocladium pyriformis sp. nov. Trans. Br. mycol. Soc. 51: 160-164.

DIXON, P. A. (1959) - Life history and cytology of Ascocybe grovesii. Ann. Bot., N.S. 23: 509-520.

DOBBS, C. G. (1938) - The life history and morphology of Dicranophora fulva. Trans. Br. mycol. Soc. 21: 167-192.

DODGE, C. W. (1929) - The higher Plectascales. Ann. mycol. 27: 145-184.

DODGE, C. W. (1935) - Medical Mycology. St. Louis, 900 pp.

DOGUET, G. (1955) - Le genre Melanospora. Botaniste 39, 313 pp.

DOGUET, G. (1956) - Le genre Thielavia. Rev. Mycol., suppl. colon. 21: 1-21.

DOGUET, G. (1956) - Morphologie et organogénie du Neocosmospora vasinfecta et N. africana. Ann. Sci. nat., Bot., Sér. 11, 17: 353-370.

DOGUET, G. (1957) - Organogénie du Microascus stysanophorus. Bull. Soc. mycol. Fr. 73: 165-178.

DOI, Y. (1966) - A revision of Hypocreales with cultural observation I. Bull. Nat. Sci. Mus. Tokyo 9: 345-357.

DOI, Y. (1968) - Revision of the Hypocreales with cultural observations II. Bull. Nat. Sci. Mus. Tokyo 11: 185-189.

DOI, Y. (1972) - Revision of the Hypocreales with cultural observations IV. The genus Hypocrea and its allies in Japan. Bull. natn. Sci. Mus. Tokyo 15: 649-751.

DOMSCH, K. H. & W. GAMS (1970) - Pilze aus Agrarböden. Stuttgart, 222 pp.

DOMSCH, K.H. & W. GAMS (1972) - Fungi in agricultural soils. Longman, London, 290 pp.

DONK, M. A. (1956) - Notes on resupinate Hymenomycetes III. Fungus 26: 3-24.

DONK, M. A. (1958) - Notes on resupinate Hymenomycetes V. Fungus 28: 16-36.

DONK, M. A. (1964) - Nomina conservanda proposita I. Proposals in Fungi. Regnum veg. 34: 7-43.

DORENBOSCH, M. M. J. (1970) - Key to nine ubiquitous soil-borne Phoma-like fungi. Persoonia 6: 1-14.

DOWNING, M. H. (1953) - Botryotrichum and Coccospora. Mycologia 45: 934-940.

DRECHSLER, C. (1923) - Some graminicolous species of Helminthosporium I. J. agric. Res. 24: 641-740.

DRECHSLER, C. (1934) - Phytopathological and taxonomic aspects of Ophiobolus, Pyrenophora, Helminthosporium and a new genus, Cochliobolus. Phytopathology 24: 953-983.

DRECHSLER, C. (1937) - Some Hyphomycetes that prey on free-living terricolous nematodes. Mycologia 29: 446-552.

DRECHSLER, C. (1947) - A Basidiobolus producing elongated secondary conidia with adhesive beaks. Bull. Torrey bot. Club 74: 403-413.

DRECHSLER, C. (1950) - Several species of Dactylella and Dactylaria that capture free-living nematodes. Mycologia 42: 1-79.

DRECHSLER, C. (1956) - Supplementary developmental stages of Basidiobolus ranarum and Basidiobolus haptosporus. Mycologia 48: 655-676.

DUBE, H. C. & K. S. BILGRAMI (1966) - Pestalotia or Pestalotiopsis. Mycopath. Mycol. appl. 29: 33-54.

DURRELL, L. W. (1959) - Some studies of Emericellopsis. Mycologia 51: 31-43.

ECKBLAD, F.-E. (1968) - The genera of the operculate Discomycetes. Nytt Mag. Bot. 15, 191 pp.

ELLIS, J. J. (1963) - A study of Rhopalomyces elegans in pure culture. Mycologia 55: 183-198.

ELLIS, J. J. & C. W. HESSELTINE (1965) - The genus Absidia: globose-spored species. Mycologia 57: 222-235.

ELLIS, M. B. (1957) - Some species of Corynespora. Mycol. Pap. *65*, 15 pp.

ELLIS, M. B. (1957) - Some species of Deightoniella. Mycol. Pap. *66*, 12 pp.

ELLIS, M. B. (1957) - Haplobasidion, Lacellinopsis and Lacellina. Mycol. Pap. *67*, 15 pp.

ELLIS, M. B. (1958/1959) - Clasterosporium and some allied Dematiaceae I and II. Mycol. Pap. *70*, 89 pp. *72*, 75 pp.

ELLIS, M. B. (1960) - Dematiaceous Hyphomycetes I. Mycol. Pap. *76*, 36 pp.

ELLIS, M. B. (1961) - Dematiaceous Hyphomycetes III. Mycol. Pap. *82*, 55 pp.

ELLIS, M. B. (1963) - Dematiaceous Hyphomycetes V. Mycol. Pap. *93*, 33 pp.

ELLIS, M. B. (1965) - Dematiaceous Hyphomycetes VI. Mycol. Pap. *103*, 46 pp.

ELLIS, M. B. (1966) - Dematiaceous Hyphomycetes VII: Curvularia, Brachysporium etc. Mycol. Pap. *106*, 57 pp.

ELLIS, M.B. (1971) - Dematiaceous Hyphomycetes. CMI., Kew, 608 pp.

ELLIS, M. B., E. A. ELLIS & J. P. ELLIS (1951) - British marsh and fen fungi. I. Trans. Br. mycol. Soc. *34*: 147-169.

EMERSON, R. (1941) - An experimental study of the life cycles and taxonomy of Allomyces. Lloydia *4*: 77-144.

EMERSON, R. & E. C. CANTINO (1948) - The isolation, growth and metabolism of Blasto-cladia in pure culture. Am. J. Bot. *35*: 157-171.

EMERSON, R. & H. C. WHISLER (1968) - Cultural studies of Oedogoniomyces and Harpochy-trium, and a proposal to place them in a new order of aquatic Phycomycetes. Arch. Mikrobiol. *61*: 195-211.

EMMONS, C. W. (1935) - The ascocarps in species of Penicillium. Mycologia *27*: 128-150.

EMMONS, C. W., C. H. BINFORD and J. P. UTZ (1963) - Medical Mycology. London, 380 pp.

ENDE, G. van den (1958) - Untersuchungen über den Pflanzenparasiten Verticillium albo-atrum. Acta bot. neerl. *7*: 665-740.

ERIKSSON, O. (1967) - On graminicolous pyrenomycetes from Fennoscandia 2. Ark. Bot. *6*: 381-440.

FALL, J. (1951) - Studies on fungus parasites of strawberry leaves in Ontario. Can. J. Bot. *29*: 299-315.

FASSATIOVÁ, O. (1967) - Notes on the genus Humicola II. Českǎ Mykol. *21*: 78-89.

FELL, J. W. & I. L. HUNTER (1968) - Isolation of heterothallic yeast strains of Metschnikowia and their mating reactions with Chlamydozyma spp. Antonie van Leeuwenhoek *34*: 365-376.

FERGUS, C. L. (1960) - A note on the occurrence of Peziza ostracoderma. Mycologia *52*: 959-961.

FERGUS, C. L. (1964) - Thermophilic and thermotolerant molds and Actinomycetes of mushroom compost during peak heating. Mycologia *56*: 267-284.

FILER, T. H. (1969) - New species of Pteridiospora. Mycologia *61*: 167-169.

FISCHER, G. W. (1953) - Manual of North American smut fungi. New York, 343 pp.

FOCKE, I. (1962) - Zwei Acremoniella-Arten als Samenbewohner von Zea Mays. Z. Pflanzenkr. *69*: 410-413.

FREDERICK, L., F. A. UECKER & C. R. BENJAMIN (1969) - A new species of Neurospora from soil of West Pakistan. Mycologia *61*: 1077-1084.

FREY, D. (1965) - Isolation of keratinophilic and other fungi from soils collected in Australia and New Guinea. Mycologia *57*: 202-215.

LE GAL, M. & F. MANGENOT (1958-1966) - Contribution à l'étude des Mollisioidées I-IV. Rev. mycol. *23*: 28-86 (1958), *25*: 135-214 (1960), *26*: 263-331 (1961), *31*: 3-44 (1966).

GAMS, W. (1966) - Zwei Arten von Chaetomium mit unregelmässig geformten Ascosporen. Nova Hedwigia *12*: 385-388.

GAMS, W. (1968) - Typisierung der Gattung Acremonium. Nova Hedwigia *16*: 141-145.

GAMS, W. (1968) - Two new species of Wardomyces. Trans. Br. mycol. Soc. *51*: 798-802.

GAMS, W. (1969) - Gliederungsprinzipien in der Gattung Mortierella. Nova Hedwigia *18*: 30-44.

GAMS, W. (1971) - Cephalosporium-artige Hyphomyceten. G. Fischer, Stuttgart, 262 pp.

GAMS, W., CHIU-YUAN CHIEN & K.H. DOMSCH (1972) - Zygospore formation by the heterothallic Mortierella elongata and a related homothallic species. Trans. Br. mycol. Soc. *58*: 5-13.

GAMS, W. & K. H. DOMSCH (1969) - Bemerkungen zu einigen schwer bestimmbaren Bodenpilzen. Nova Hedwigia *18*: 1-29.

GAMS, W. & M. GERLAGH (1968) - Beiträge zur Systematik und Biologie von Plectosphaerella cucumeris und der zugehörigen Konidienform. Persoonia *5*: 177-188.

GAMS, W. & A. C. M. HOOZEMANS (1970) - Cladobotryum-Konidienformen von Hypomyces-Arten. Persoonia *6*: 95-110.

GAMS, W. & S. T. WILLIAMS (1963) - Heterothallism in Mortierella parvispora. Nova Hedwigia *5*: 347-357.

GAERTNER, A. (1954) - Über das Vorkommen einiger niederer Erdphycomyceten in Afrika, Schweden und an einigen mitteleuropäischen Standorten. Arch. Mikrobiol. *21*: 4-56.

GEORG, L. K., L. AJELLO, L. FRIEDMANN & S. A. BRINKMAN (1962) - A new species of Microsporum pathogenic to man and animals. Sabouraudia *1*: 189-196.

GERLACH, W. (1956) - Beiträge zur Kenntnis der Gattung Cylindrocarpon I. Phytopath. Z. *26*: 161-170.

GERLACH, W. (1959) - Beiträge zur Kenntnis der Gattung Cylindrocarpon III. Phytopath. Z. *35*: 333-346.

GERLACH, W. (1968) - Calonectria uniseptata n. sp., die bisher unbekannte Hauptfruchtform von Cylindrocladium scoparium. Phytopath. Z. *61*: 372-381.

GERLACH, W. (1970) - Suggestions to an acceptable modern Fusarium system. Annls Acad. Sci. fenn. A, IV, Biologica *168*: 37-49.

GERLACH, W. & D. ERSHAD (1970) - Beitrag zur Kenntnis der Fusarium- und Cylindrocarpon-Arten in Iran. Nova Hedwigia *20*: 725-784.

GILMAN, J. C. (1957) - A Manual of Soil Fungi. Ed. 2. Iowa State College Press, Ames, Iowa, 450 pp.

GLENN—BOTT, J. I. (1955) - On Helicodendron tubulosum and some similar species. Trans. Br. mycol. Soc. *38*: 17-30.

GLICK, A.D. & K.J. KWON-CHUNG (1973) - Ultrastructural comparison of coils and ascospores of Emmonsiella capsulata and Ajellomyces dermatitidis. Mycologia *65*: 216-220.

GOOS, R. D. (1964) - A new record of Cephaliophora irregularis. Mycologia *56*: 133-136.

GOOS, R. D. (1967) - Observations on Riessia semiophora. Mycologia *59*: 718-722.

GOOS, R.D. (1969) - Conidium ontogeny in Cacumisporium capitulatum. Mycologia *61*: 52-56.

GOOS, R. D., E. A. COX & G. STOTZKY (1961) - Botryodiplodia theobromae and its association with Musa species. Mycologia *53*: 262-277.

GORDON, W. L. (1952) - The occurrence of Fusarium species in Canada. II. Can. J. Bot. *30*: 209-251.

GORDON, W.L. (1956) - The taxonomy and habitats of the Fusarium species in Trinidad. Can. J. Bot. *34*: 847-864.

GRAAFLAND, W. (1953) - Four species of Exobasidium in pure culture. Acta bot. neerl. *1*: 516-522.

GRAAFLAND, W. (1960) - The parasitism of Exobasidium japonicum on Azalea. Acta bot. neerl. *9*: 347-379.

GRAHAM, J. H. & E. S. LUTTRELL (1961) - Species of Leptosphaerulina on forage plants. Phytopathology *51*: 650-693.

GRANITI, A. (1962) - Scolecobasidium anelli n. sp. agente di annerimenti superficiali di stalattiti. G. bot. ital. *69*: 360-365.

GREMMEN, J. (1954) - Taxonomical notes on mollisiaceous fungi I. Fungus *24*: 1-8.

GREMMEN, J. (1955) - Taxonomical notes on mollisiaceous fungi II. Fungus *25*: 1-12.

GREMMEN, J. (1956) - Taxonomical notes on mollisiaceous fungi III and IV. Fungus *26*: 28-37.

GREMMEN, J. (1965) - Poplar-inhabiting Drepanopeziza species and their life-history. Nova Hedwigia *9*: 170-176.

GRIFFIN, H. D. (1968) - The genus Ceratocystis in Ontario. Can. J. Bot. *46*: 689-718.

GROSKLAGS, J. H. & M. E. SWIFT (1957) - The perfect stage of an antibiotic-producing Cephalosporium. Mycologia *49*: 305-317.

GROVE, W. B. (1935) - British Stem- and Leaf-fungi. 1: Sphaeropsidales. Cambridge, 488 pp.

GROVE, W. B. (1937) - British Stem- and Leaf-fungi. 2: Sphaeropsidales and Melanconiales. Cambridge, 406 pp.

GROVES, J. W. (1946) - North American species of Dermea. Mycologia *38*: 351-431.

GROVES, J. W. (1952) - The genus Tympanis. Can. J. Bot. *30*: 571-651.

GROVES, J. W. (1965) - The genus Godronia. Can. J. Bot. *43*: 1195-1276.

GROVES, J. W. & A. J. SKOLKO (1944) - Notes on seed-borne fungi. I. Stemphylium. Can. J. Res., Sect. C, *22*: 190-199.

GROVES, J. W. & A. J. SKOLKO (1964) - Notes on seed-borne fungi. IV. Acremoniella, Chlamydomyces and Trichocladium. Can. J. Res., Sect. C, *24*: 74-80.

GUBA, E. F. (1961) - Monograph of Monochaetia and Pestalotia. Harvard Univ. Press, Cambridge, Mass., 342 pp.

GUILLIERMOND, A. (1936) - L'Eremothecium ashbyii, nouveau champignon parasite des capsules du cotonnier. Rev. Mycol. *1*: 115-156.

GUNNERBECK, E. (1971) - Studies in foliicolous Deuteromycetes I. The genus Mastigosporium in Sweden. Svensk bot. Tidskr. *65*: 39-52.

GUSTAFSSON, M. (1965) - On species of the genus Entomophthora in Sweden. I. Classification and Distribution. Lantbr.-Högsk. Annlr *31*: 102-212.

HAARD, K. (1968) - Taxonomic studies on the genus Arthrobotrys. Mycologia *60*: 1140-1159.

HADLOK, R. & A. C. STOLK (1969) - Eurotium leucocarpum sp. n. Antonie van Leeuwenhoek *35*: 9-12.

HAHN, G. G. (1957) - A new species of Phacidiella causing the so-called Phomopsis disease on conifers. Mycologia *49*: 226-239.

HANLIN, R. T. (1961) - Studies in the genus Nectria - I. Factors influencing perithecial formation in culture. Bull. Torrey bot. Club *88*: 95-103.

HANLIN, R. T. (1961) - Studies in the genus Nectria - II. Morphology of N. gliocladioides. Am. J. Bot. *48*: 900-908.

HANLIN, R. T. (1963) - Morphology of Neuronectria peziza. Am. J. Bot. *50*: 56-66.

HANLIN, R. T. (1963) - Morphology of Hypomyces lactifluorum. Bot. Gaz. *124*: 395-401.

HANLIN, R. T. (1964) - Morphology of Hypomyces trichothecioides. Am. J. Bot. *51*: 201-208.

HANSFORD, C. G. (1946) - The foliicolous Ascomycetes, their parasites and associated fungi. Mycol. Pap. *15*, 240 pp.

HARROLD, C. E. (1950) - Studies in the genus Eremascus. Ann. Bot. *14*: 127-148.

HASEGAWA, T. (1965) - A report on the taxonomy of red to orange Rhodotorula. Ann. Rep. Inst. Ferment. Osaka *2*: 1-25.

HAWKSWORTH, D.L. (1971) - A revision of the genus Ascotricha. Mycol. Pap. *126*, 28 pp.

HEJAROUDE, G.-A. (1969) - Études taxonomiques sur les Phaeosphaeria et leurs formes voisines. Sydowia *22*: 57-107.

HENNEBERT, G. L. (1962) - Wardomyces and Asteromyces. Can. J. Bot. *40*: 1203-1216.

HENNEBERT, G. L. (1963) - Les Botrytis des Allium, Meded. Landb. Hogesch. OpzoekStns. Gent *28*: 851-876.

HENNEBERT, G. L. (1964) - Botryotinia squamosa, nouveau parasite de l'oignon en Belgique. Parasitica *20*: 138-153.

HENNEBERT, G. L. (1967) - Chalaropsis punctulata, a new hyphomycete. Antonie van Leeuwenhoek *33*: 333-340.

HENNEBERT, G. L. (1968) - New species of Spirosphaera. Trans. Br. mycol. Soc. *51*: 13-24.

HENNEBERT, G. L. (1968) - Echinobotryum, Wardomyces and Mammaria. Trans. Br. mycol. Soc. *51*: 749-762.

HENNEBERT, G. L. & J. W. GROVES (1963) - Three new species of Botryotinia on Ranunculaceae. Can. J. Bot. *41*: 341-370.

HERING, T. F. (1965) - British records 88. Trans. Br. mycol. Soc. *48*: 663-667.

HESSELTINE, C. W. (1953) - A revision of the Choanephoraceae. Am. Midl. Nat. *50*: 248-256.

HESSELTINE, C. W. (1954) - The section genevensis of the genus Mucor. Mycologia *46*: 358-366.

HESSELTINE, C. W. (1955) - Genera of the Mucorales with notes on their synonymy. Mycologia *47*: 344-363.

HESSELTINE, C. W. (1958) - The genus Syzygites. Lloydia *20*: 228-237.

HESSELTINE, C. W. (1960) - Gilbertella gen. nov. (Mucorales). Bull. Torrey bot. Club *87*: 21-30.

HESSELTINE, C. W. & P. ANDERSON (1956) - The genus Thamnidium and a study of the formation of its zygospores. Am. J. Bot. *43*: 696-702.

HESSELTINE, C. W. & P. ANDERSON (1957) - Two genera of molds with low temperature growth requirements. Bull. Torrey bot. Club *84*: 31-45.

HESSELTINE, C. W. & C. R. BENJAMIN (1957) - Notes on the Choanephoraceae. Mycologia *49*: 723-733.

HESSELTINE, C. W., C. R. BENJAMIN & B. S. MEHROTRA (1959) - The genus Zygorhynchus. Mycologia *51*: 173-194.

HESSELTINE, C. W. & J. J. ELLIS (1961) - Notes on Mucorales, especially Absidia. Mycologia *53*: 406-426.

HESSELTINE, C. W. & J. J. ELLIS (1964) - An interesting species of Mucor, M. ramosissimus. Sabouraudia *3*: 151-154.

HESSELTINE, C. W. & J. J. ELLIS (1964) - The genus Absidia, Gongronella and cylindrical-spored species of Absidia. Mycologia *56*: 568-601.

HESSELTINE, C. W. & J. J. ELLIS (1966) - Species of Absidia with ovoid sporangiospores. Mycologia *58*: 761-785.

HESSELTINE, C. W. & D. I. FENNELL (1955) - The genus Circinella. Mycologia *47*: 193-212.

HINDS, T. E. & R. W. DAVIDSON (1967) - A new species of Ceratocystis on aspen. Mycologia *59*: 1102-1106.

HÖHNK, W. (1952) - Die in Nordwestdeutschland gefundenen ufer- und bodenbewohnenden Saprolegniaceae. Veröff. Inst. Meeresforsch. Bremerhaven *1*: 52-90, 115-125, 126-128, 247-278.

HÖHNK, W. (1953) - Studien zur Brack- und Seewassermykologie III. Veröff. Inst. Meeresforsch. Bremerhaven *2*: 52-108.

HOLM, L. (1948) - The Swedish species of the genus Ophiobolus. Svensk bot. Tidskr. *42*: 337-347.

HOLM, L. (1952) - The herbicolous Swedish species of the genus Leptosphaeria. Svensk bot. Tidskr. 46: 18-46.

HOLM, L. (1957) - Etudes taxonomiques sur les Pléosporacées. Symb. bot. uppsal. 14, (3),188 pp.

HOLUBOVÁ–JECHOVÁ, V. (1969) - New species of the genus Oidium. Česká Mykol. 23: 209-221.

HONEY, E. E. (1928) - The monilioid species of Sclerotinia. Mycologia 20: 127-156.

HONEY, E. E. (1936) - North American species of Monilinia I. Occurrence, grouping and life histories. Am. J. Bot. 23: 100-106.

HOOG, G.S. de (1972) - The genera Beauveria, Isaria, Tritirachium and Acrodontium. Stud. Mycol. 1, 41 pp.

HOOG, G.S. de & P.J. MULLER (1973) - A new species of Embellisia associated with skin disease of hyacinths. Neth. J. Pl. Path. 79: 85-93.

HORENSTEIN, E. A. & E. C. CANTINO (1961) - Morphogenesis in and the effect of light on Blastocladiella britannica sp. nov. Trans. Br. mycol. Soc. 44: 185-198.

HOTSON, H. H. (1942) - Some species of Papulaspora associated with rots of Gladiolus bulbs. Mycologia 34: 391-398.

HOTSON, J. W. (1917) - Notes on bulbiferous fungi with a key to described species. Bot. Gaz. 64: 265-284.

HOWELL, A. (1939) - Studies on Histoplasma capsulatum and similar form species. I. Morphology and development. Mycologia 31: 191-216.

HUBERT, E. E. (1935) - Observations on Tuberculina maxima, a parasite of Cronartium ribicola. Phytopathology 25: 253-261.

HUDSON, H. J. (1963) - Pyrenomycetes of sugar cane and other grasses in Jamaica. II. Conidia of Apiospora montagnei. Trans. Br. mycol. Soc. 46: 19-23.

HUDSON, H. J. (1963) - The perfect state of Nigrospora oryzae. Trans. Br. mycol. Soc. 46: 355-360.

HUGHES, S. J. (1949) - Studies on some diseases of Sainfoin (Onobrychis sativa) II. Trans. Br. mycol. Soc. 32: 34-59.

HUGHES, S. J. (1949) - Studies on Microfungi II. Mycol. Pap. 31, 33 pp.

HUGHES, S. J. (1950) - Studies on Microfungi IV. Mycol. Pap. 37, 17 pp.

HUGHES, S. J. (1951) - Studies on Microfungi IX, X, XI, XII. Mycol. Pap. 43, 25 pp; 44, 18 pp; 45, 36 pp; 46, 35 pp.

HUGHES, S. J. (1952) - Four species of Septonema. Naturalist, Hull p. 7-12.

HUGHES, S. J. (1952) - Trichocladium Harz. Trans. Br. mycol. Soc. 35: 152-157.

HUGHES, S. J. (1952) - Speira stipitata. Trans. Br. mycol. Soc. 35: 243-247.

HUGHES, S. J. (1953) - Some foliicolous Hyphomycetes. Can. J. Bot. 31: 560-576.

HUGHES, S. J. (1953) - Conidiophores, conidia and classification. Can. J. Bot. 31: 577-659.

HUGHES, S. J. (1955) - Microfungi I. Cordana, Brachysporium, Phragmocephala. Can. J. Bot. 33: 259-268.

HUGHES, S. J. (1957) - Microfungi III. Mammaria. Sydowia, Beiheft 1: 359-363.

HUGHES, S. J. (1958) - Revisiones Hyphomycetum aliquot cum appendice de nominibus rejiciendis. Can. J. Bot. 36: 727-836.

HUGHES, S. J. (1959) - Microfungi IV. Trichocladium canadense n. sp. Can. J. Bot. 37: 857-859.

HUGHES, S. J. (1960) - Microfungi V. Conoplea and Exosporium. Can. J. Bot. 38: 659-696.

HUGHES, S. J. (1965) - New Zealand Fungi. 3. Catenularia Grove. N. Z. Jl Bot. 3: 136-150.

HUGHES, S. J. & W. B. KENDRICK (1963) - Microfungi IX. Menispora. Can. J. Bot. 41: 693-718.

HUGHES, S. J. & W. B. KENDRICK (1968) - New Zealand Fungi 12. Menispora, Codinaea, Menisporopsis. N. Z. Jl Bot. *6*: 323-375.

HUNT, J. (1956) - Taxonomy of the genus Ceratocystis. Lloydia *19*: 1-58.

HÜTTER, R. (1958) - Untersuchungen über die Gattung Pyrenopeziza. Pytopath. Z. *33*: 1-54.

INGOLD, C. T. (1942) - Aquatic Hyphomycetes of decaying alder leaves. Trans. Br. mycol. Soc. *25*: 339-417.

INGOLD, C. T. (1953) - Dispersal in Fungi. Oxford, 197 pp.

INGOLD, C. T. (1956) - The conidial apparatus of Trichothecium roseum. Trans. Br. mycol. Soc. *39*: 460-464.

INGOLD, C. T. (1958) - Aquatic Hyphomycetes from Uganda and Rhodesia. Trans. Br. mycol. Soc. *41*: 109-114, 365-372.

INGOLD, C. T. & V. J. COX (1957) - On Tripospermum and Campylospora. Trans. Br. mycol. Soc. *40*: 317-321.

INGOLD, C. T., V. DANN & P. McDOUGALL (1968) - Tripospermum camelopardus sp. nov. Trans. Br. mycol. Soc. *51*: 51-56.

INGOLD, C. T., P. McDOUGALL & V. DANN (1968) - Volucrispora graminea sp. nov. Trans. Br. mycol. Soc. *51*: 325-328.

INUI, T., Y. TAKEDA & H. IIZUKA (1965) - Taxonomical studies on genus Rhizopus. J. gen. appl. Microbiol. *11*, suppl., 121 pp.

ISAAC, I. (1949) - A comparative study of pathogenic isolates of Verticillium. Trans. Br. mycol. Soc. *32*: 137-157.

ISAAC, I. (1953) - A further comparative study of pathogenic isolates of Verticillium. Trans. Br. mycol. Soc. *36*: 180-195.

ISAAC, I. (1967) - Speciation in Verticillium. A. Rev. Phytopath. *5*: 201-222.

ISAAC, I. & R. R. DAVIES (1955) - A new hyaline species of Verticillium: V. intertextum sp. nov. Trans. Br. mycol. Soc. *38*: 143-156.

JACKSON, A. E. & E. R. DEARDEN (1948) - Martensella corticii Thaxter and its distribution. Mycologia *40*: 168-176.

JACQUES, J. E. (1941) - Studies in the genus Heterosporium. Contr. Inst. Bot. Montreal, *39*, 46 pp.

JENKINS, A. E. & A. A. BITANCOURT (1941) - Revised descriptions of the genera Elsinoë and Sphaceloma. Mycologia *33*: 338-340.

JENKINS, A. E. & A. A. BITANCOURT (1956) - Myriangiales selecti exsiccati, fascicle 1-10. Rev. Soc. bras. Argron. *12*: 41-68.

JENKINS, A. E. & A. A. BITANCOURT (1966) - Myriangiales selecti exsiccati, fascicle 11. Anais Acad. bras. Cienc. *38*: 525-536.

JENSEN, C. N. (1912) - Fungus flora of the soil. Bull. Cornell agric. Exp. Stn *315*: 414-501.

JOHANNES, H. (1955) - Die Gattung Cladolegnia gen. nov. (Isoachlya Kauffman) der Saprolegniaceae Nees. Feddes Rep. *58*: 209-220.

JOHNSON, T. W. (1956) - The genus Achlya: Morphology and Taxonomy. Univ. Mich. Press, Ann Arbor, 180 pp.

JOHNSON, T. W. & J. SURRATT (1955) - Taxonomy of the species of Isoachlya possessing single oospores. Mycologia *47*: 122-129.

JOLY, P. (1964) - Le genre Alternaria. Encycl. mycol. *33*, 250 pp.

JOLY, P. (1964) - Recherches sur la nature et le mode de formation des spores chez Torula. Bull. Soc. mycol. Fr. *80*: 186-196.

JONG, S.C. & E.E. DAVIS (1972) - Phialocephala humicola a new hyphomycete. Mycologia *64*: 1351-1356.

JØRSTAD, I. (1967) - Septoria and septorioid fungi on Gramineae in Norway. Skr. Nor. Vid. Akad. Oslo I. mat.-nat. Kl. *24*, 63 pp.

JUEL, H. O. (1925) - Mykologische Beiträge IX. Ark. Bot. *19* (20), 1-10.

KAMAT, M. N. & V. G. RAO (1970) - The genus Curvularia from India. Nova Hedwigia *18*: 597-626.

KANOUSE, B.B. (1958) - Some species of the genus Trichophaea. Mycologia *50*: 121-140.

KARLING, J. S. (1942) - The Plasmodiophorales. New York, 144 pp.

KARLING, J. S. (1964) - Synchytrium. New York, 470 pp.

KEENER, P. D. (1934) - Biological specialization in Darluca filum. Bull. Torrey bot. Club *61*: 475-490.

KENDRICK, W. B. (1958) - Sympodiella, a new hyphomycetous genus. Trans. Br. mycol. Soc. *41*: 519-521.

KENDRICK, W. B. (1958) - Helicoma monospora from pine litter. Trans. Br. mycol. Soc. *41*: 446-448.

KENDRICK, W. B. (1962) - The Leptographium complex. Verticicladiella. Can. J. Bot. *40*: 771-797.

KENDRICK, W. B. (1963) - The Leptographium complex. Two new species of Phialocephala. Can. J. Bot. *41*: 1015-1023.

KENDRICK, W. B. (1964) - The Leptographium complex. Hantzschia Auerswald. Can. J. Bot. *42*: 1291-1295.

KENDRICK, W.B. (Ed.) (1971) - Taxonomy of Fungi Imperfecti. Univ. Toronto Press, 309 pp.

KENDRICK, W. B. & G. C. BHATT (1966) - Trichocladium opacum. Can. J. Bot. *44*: 1728-1730.

KENDRICK, W.B. & J.W. CARMICHAEL (1973) - Hyphomycetes. In: The Fungi, vol. IV A: 323-509.

KENDRICK, W. B., G. T. COLE & G. C. BHATT (1968) - Conidium ontogeny in hyphomycetes. Gonatobotryum apiculatum and its botryose blastospores. Can. J. Bot. *46*: 591-596.

KENDRICK, W. B. & A. C. MOLNAR (1965) - A new Ceratocystis and its Verticicladiella imperfect state associated with the bark beetle Dryocoetes confusus on Abies lasiocarpa. Can. J. Bot. *43*: 39-43.

KIMBROUGH, J. W. (1966) - Studies in the Pseudoascoboleae. Can. J. Bot. *44*: 685-704.

KIMBROUGH, J. W. (1969) - North American species of Thecotheus. Mycologia *61*: 99-114.

KIMBROUGH, J. W. & R. P. KORF (1967) - A synopsis of the genera and species of the tribe Teleboleae (= Pseudoascoboleae). Am. J. Bot. *54*: 9-29.

KIMBROUGH, J. W., E. R. LUCK—ALLEN & R. F. CAIN (1969) - Iodophanus, the Pezizeae segregate of Ascophanus (Pezizales). Am. J. Bot. *56*: 1187-1202.

KLEBAHN, H. (1907) - Untersuchungen über einige Fungi imperfecti und die zugehörigen Ascomycetenformen. Z. PflKrankh. PflPath. PflSchutz *17*: 223-237.

KLEBAHN, H. (1918) - Haupt- und Nebenfruchtformen der Ascomyceten. Leipzig, 395 pp.

KLINGSTRÖM, A. & L. BEYER (1965) - Two new species of Scytalidium with antagonistic properties to Fomes annosus. Svensk bot. Tidskr. *59*: 30-36.

KOBAYASHI, T. (1968) - Notes on Japanese species of the genus Melanconium. Trans. mycol. Soc. Japan *9*: 1-11.

KOBAYASHI, T. (1970) - Taxonomic studies of Japanese Diaporthaceae with special reference to their life-histories. Bull. Gov. Forest Exp. Stn *226*, 242 pp.

KOBAYASHI, T. & K. ITO (1956) - Notes on the genus Endothia in Japan. I. Bull. Gov. For. Exp. Stn, Tokyo *92*: 81-97.

KORF, R. P. (1960) - Nomenclatural notes IV. The generic name Plicaria. Mycologia *52*: 648-651.

KORF, R.P. (1973) - Discomycetes and Tuberales. In The Fungi, vol. IV A: 249-319.

KRAMER, C. L. (1961) - Morphological development and nuclear behaviour in the genus Taphrina. Mycologia 52: 295-320.

KRAUSE, R.A. & R.K. WEBSTER (1972) - The morphology, taxonomy and sexuality of the rice stem rot fungus, Magnaporthe salvinii (Leptosphaeria salvinii). Mycologia 64: 103-114.

KREGER—VAN RIJ, N. J. W. (1964) - A taxonomic study of the yeast genera Endomycopsis, Pichia and Debaryomyces. Groningen, 194 pp.

KREGER—VAN RIJ, N. J. W. & D. G. AHEARN (1968) - Shape and structure of the ascospores of Hanseniaspora uvarum. Mycologia 60: 604-612.

KRENNER, J. A. (1961) - Studies in the field of the microscopic fungi. III. On Entomophthora aphidis with special regard to the family of the Entomophthoraceae in general. Acta bot. hung. 7: 345-376.

KRNETA—JORDI, M. (1962) - Cytologische und physiologische Untersuchungen an Eremothecium ashbyii. Arch. Mikrobiol. 43: 76-108.

KUDRJAWZEW, W. I. (1960) - Die Systematik der Hefen. Berlin, 324 pp. (German translation of the Russian edition, 1954).

KUEHN, H. H. (1960) - Observations on Gymnoascaceae. VIII. A new species of Arthroderma. Mycopath. Mycol. appl. 13: 189-197.

KUEHN, H. H., G. F. ORR & G. R. GHOSH (1964) - Pathological implications of Gymnoascaceae. Mycopath. Mycol. appl. 24: 35-46.

KULIK, M. M. (1968) - A compilation of descriptions of new Penicillium species. U.S.Dep. Agric. Handbook 351, 80 pp.

KWON-CHUNG, K.J. (1973) - Studies on Emmonsiella capsulata I. Heterothallism and development of the ascocarp. Mycologia 65: 109-121.

LAFFIN, R. J. & V. M. CUTTER (1959) - Investigations on the life cycle of Sporidiobolus johnsonii. J. Elisha Mitchell scient. Soc. 75: 89-100.

LAKON, G. (1963) - Entomophthoraceae. Nova Hedwigia 5: 6-26.

LANJOUW, J. & al. (1966) - International Code of Botanical Nomenclature. Utrecht.

LANGERON, M. (1949) - Remarques sur les genres Beauveria et Tritirachium. Rev. Mycol. 14: 133-136.

LATCH, G. C. M. (1965) - Metarrhizium anisopliae (Metschnikoff) Sorokin strains in New Zealand and their possible use for controlling pasture inhabiting insects. N. Z. Jl agric. Res. 8: 384-386.

LENTZ, P. L. (1966) - Dactylaria in relation to the conservation of Dactylium. Mycologia 58: 965-966.

LIMBER, D. P. (1955) - Studies in the genus Sporonema. Mycologia 47: 344-363.

LIMBER, D. P. & E. K. CASH (1945) - Actinopelte dryina. Mycologia 37: 129-137.

LINDAU, G. (1907 and 1910) - Hyphomycetes in Rabenhorst Krypt.Fl., Pilze VIII & IX, Leipzig, 852 & 983 pp.

LINDEBERG, B. (1959) - Ustilaginales of Sweden. Symb. bot. uppsal. 16, 2, 175 pp.

LINDER, D. H. (1929) - A monograph of the helicosporous Fungi Imperfecti. Ann. Mo. bot. Gdn 16: 227-388.

LINDER, D. H. (1942) - A contribution towards a monograph of the genus Oidium. Lloydia 5: 165-207.

LINDER, D. H. (1943) - The genera Kickxella, Martensella and Coemansia. Farlowia 1: 49-77.

LINDER, D. H., in BARGHOORN, E. S. and D. H. LINDER (1944) - Marine fungi, their taxonomy and biology. Farlowia 1: 395-467.

LINNEMANN, G. (1941) - Die Mucorineen-Gattung Mortierella. Pflanzenforschung 23, 64 pp.

286

LINNEMANN, G. (1958) - Untersuchungen über die Verbreitung und Systematik der Mortierellen. Arch. Mikrobiol. *30*: 256-267.

LITVINOV, M. A. (1967) - Key to microscopic soil fungi. (in Russian) Leningrad, 303 pp., 178 figs.

LODDER, J. & N. J. W. KREGER—VAN RIJ (1952) - The Yeasts. Amsterdam, 713 pp.

LODDER, J. (edit.) (1970) - The yeasts, a taxonomic study. Amsterdam, 1385 pp.

LODHA, B. C. (1963) - Notes on two species of Trichurus. J. Indian bot. Soc. *42*: 135-142.

LODHA, B. C. (1964) - Studies on coprophilous fungi. I. Chaetomium. J. Indian bot. Soc. *43*: 121-140.

LODHA, B. C. (1964) - Studies on coprophilous fungi. II. Chaetomium. Antonie van Leeuwenhoek *20*: 163-167.

LOHMAN, M. A. (1942) - A new fungous parasite on dung-inhabiting Ascomycetes. Mycologia *34*: 104-111.

LUCAS, G. B. (1949) - Studies on the morphology and cytology of Thielavia basicola. Mycologia *41*: 553-560.

LUNDQVIST, N. (1964) - Zygopleurage and Zygospermella. Bot. Notiser *122*: 353-374.

LUNDQVIST, N. (1972) - Nordic Sordariaceae. Symb. bot. upsal. *20*(1), 374 pp.

LUTTRELL, E. S. (1940) - Tar spot of American holly. Bull. Torrey bot. Club *67*: 692-704.

LUTTRELL, E. S. (1954) - An undescribed species of Piricularia on sedges. Mycologia *46*: 810-814.

LUTTRELL, E. S. (1963) - Taxonomic criteria in Helminthosporium. Mycologia *55*: 643-674.

LUTTRELL, E. S. (1964) - Systematics of Helminthosporium and related genera. Mycologia *56*: 119-132.

LYTHGOE, J. N. (1958) - Taxonomic notes on the genera Helicostylum and Chaetostylum (Mucoraceae). Trans. Br. mycol. Soc. *41*: 135-141.

MacGARVIE, Q. D. (1968) - Hyphomycetes on Juncus effusus I. Scient. Proc. R. Dublin Soc., B, *2*: 153-161.

MACIEJOWSKA, Z. & E. B. WILLIAMS (1963) - Studies on morphological forms of Staphylotrichum coccosporum. Mycologia *55*: 221-225.

MacLEOD, D. M. (1954) - Investigations in the genera Beauveria and Tritirachium. Can. J. Bot. *32*: 818-890.

MAHONEY, D. P., L. H. HUANG & M. P. BACKUS (1969) - New homothallic Neurosporas from tropical soils. Mycologia *61*: 264-272.

MAINS, E.B. (1960) - Species of Aschersonia (Sphaeropsidales). Lloydia *22*: 215-221.

MAIRE, R. (1903) - Remarques taxonomiques et cytologiques sur le Botryosporium pulchellum. Ann. mycol. *1*: 335-340.

MÄKILÄ, K. (1970) - The genus Mastigosporium in Finland. Karstenia *11*: 5-22.

MALLOCH, D. (1970) - New concepts in the Microascaceae illustrated by two new species. Mycologia *62*: 727-740.

MALLOCH, D. & R.F. CAIN (1970) - The genus Arachnomyces. Can. J. Bot. *48*: 839-845.

MALLOCH, D. & R.F. CAIN (1971) - New genera of Onygenaceae. Can. J. Bot. *49*: 839-846.

MALLOCH, D. & R.F. CAIN (1971) - The genus Kernia. Can. J. Bot. *49*: 855-867.

MALLOCH, D. & R.F. CAIN (1972) - New species and combinations of cleistothecial Ascomycetes. Can. J. Bot. *50*: 61-72.

MARASUS, W. F. O., G. C. A. van der WESTHUIZEN, K. T. van WARMELO & M. C. PAPENDORF (1966) - New and interesting records of South African fungi. Bothalia *9*: 229-243.

MARIAT, F., P. LAVALLE & P. DESTOMBES (1962) - Recherches sur la Sporotrichose. Sabouraudia *2*: 60-79.

MARTIN, G.W. (1941) - On Argynna polyhedron. J. Wash. Acad. Sci. *31*: 309-311.

MARTIN, P. (1967) - Studies in the Xylariaceae I-IV. Jl S. Afr. Bot. *33*: 205-240, 315-328; *34*: 153-199, 303-330.

MARVANOVÁ, L., P. MARVAN & J. RŮŽIČKA (1967) - Gyoerffyella Kol 1928, a genus of the Hyphomycetes. Persoonia *5*: 29-44.

MASON, E. W. (1933) - Annotated account of fungi received at the Imperial Bureau of Mycology, List II (fasc. 2), Mycol. Pap. *3*, 67 pp.

MASON, E. W. (1937) - Annotated account of fungi received at the Imperial Mycological Institute, List II, (fasc. 3), Mycol. Pap. *4*, 31 pp.

MASON, E. W. (1941) - Annotated account of fungi received at the Imperial Mycological Institute, List II (fasc. 3), Mycol. Pap. *5*, 101-144.

MASON, E. W. & M. B. ELLIS (1953) - British species of Periconia. Mycol. Pap. *56*, 127 pp.

MATSUSHIMA, T. (1971) - Microfungi of the Solomon Islands and Papua-New Guinea. Kobe, 78 pp., 169 figs., 48 plates.

MATTHEWS, V. D. (1931) - Studies on the genus Pythium. Chapel Hill, 136 pp.

MAZZUCCHETTI, G. (1965) - Il genere Chaetomium. Rome, 364 pp.

McKEEN, C. G. (1952) - Studies of Canadian Thelephoraceae IX. Can. J. Bot. *30*: 764-787.

MEHROTRA, B.S. & K. NAND (1967) - Studies of Mucor from India. Sydowia *20*: 67-73.

MEHROTRA, B.S. & R. PRASAD (1965) - Species of Syncephalis from India I. Sydowia *19*: 112-116.

MEHROTRA, B.S. & R. PRASAD (1967) - Species of Syncephalis from India II. Mycopath. Mycol. appl. *32*: 199-204.

MELNIK, V. (1967) - O sistematike roda Spermospora. Mikol. Fitopatol. *1*: 253-256.

MENNA, M. E. di (1965) - Yeasts in New Zealand soils. N. Z. Jl Bot. *3*: 194-203.

MENNA, M. E. di (1965) - Schizoblastosporion starkeyi-henricii. Mycopath. Mycol. appl. *25*: 205-212.

MENON, R. (1956) - Studies on Venturiaceae on rosaceous plants. Phytopath. Z. *27*: 117-146.

MENZINGER, W. (1965) - Karyologische Untersuchungen an Arten und Formen der Gattung Botrytis. Arch. Mikrobiol. *52*: 178-196.

MEYER, J. (1959) - Moisissures du sol et des litières de la région de Yangambi (Congo Belge). Publ. Inst. Nat. Étude agron. Congo Belge *75*, 211 pp.

MIDDLETON, J. T. (1943) - The taxonomy, host range and geographic distribution of the genus Pythium. Mem. Torrey bot. Club *20*, 171 pp.

MILKO, A. A. (1967) - Taxonomy and synonyms of the Mucorales. Mikol. Fitopatol. *1*: 26-35.

MILKO, A.A. (1970) - Genus Absidia (Mucorales). Novosti Sist. nizsh. Rast. 7: 121-138. (in Russian).

MILKO, A. A. & L. A. BELJAKOVA (1967) - The genus Cunninghamella and taxonomy of the Cunninghamellaceae. Mikrobiologiya *36*: 684-699.

MILLER, M. W., E. R. BARKER & J. I. PITT (1967) - Ascospore numbers in Metschnikowia. J. Bact. *94*: 258-259.

MILLER, M. W. & H. J. PHAFF (1958) - A comparative study of the apiculate yeasts. Mycopath. Mycol. appl. *10*: 113-141.

MILLER, M. W. & H. J. PHAFF (1958) - On the cell wall composition of the apiculate yeasts. Antonie van Leeuwenhoek *24*: 225-238.

MIRZA, F. (1968) - Taxonomic investigations on the ascomycetous genus Cucurbitaria. Nova Hedwigia *16*: 161-213.

MIRZA, J. H. & R. F. CAIN (1969) - Revision of the genus Podospora. Can. J. Bot. *47*: 1999-2048.

MISRA, P. C. (1967) - Torula terrestris n. sp. from soil. Can. J. Bot. *45*: 367-369.

MIX, A. J. (1949) - A monograph of the genus Taphrina. Sci. Bull. Univ. Kans. *33*: 1-167.

MOORE, E. J. (1962) - The ontogeny of the sclerotia of Pyrenoma domesticum. Mycologia *54*: 312-316.

MOORE, E. J. (1963) - The ontogeny of the apothecia of Pyrenoma domesticum. Am. J. Bot. *50*: 37-44.

MOORE, E. J. & R. P. KORF (1963) - The genus Pyrenoma. Bull. Torrey bot. Club *90*: 33-42.

MOORE, R. T. (1955, 1957) - Index to the Helicosporae. Mycologia *47*: 90-103, and Mycologia *49*: 580-587.

MOORE, R. T. (1958) - Deuteromycetes I: The Sporidesmium complex. Mycologia *50*: 681-692.

MOREAU, C. (1953) - Coexistence des formes Thielaviopsis et Graphium chez une souche de Ceratocystis major. Rev. Mycol., suppl. colon. *17*: 17-25.

MOREAU, C. (1954) - Les genres Sordaria et Pleurage. Encycl. mycol. *25*, Paris, 330 pp.

MOREAU, C. (1963) - Morphologie comparée de quelques Phialophora et variations du P. cinerescens. Rev. Mycol. *28*: 260-276.

MOREAU, C. & M. (1963) - Deux curiosités mycologiques polluant l'atmosphère d'installations industrielles. Bull. Soc. mycol. Fr. *79*: 242-248.

MOREAU–FROMENT, M. (1956) - Les Neurospora. Bull. Soc. bot. Fr. *103*: 678-738.

MORENZ, J. (1963) - Geotrichum candidum. Mykol. Schriftenreihe 1. Leipzig, 79 pp.

MORENZ, J. (1964) - Taxonomische Untersuchungen zur Gattung Geotrichum. Mykol. Schriftenreihe *2*: 33-64.

MORGAN-JONES, G. (1973) - Genera coelomycetarum VII. Cryptocline. Can. J. Bot. *51*: 309-325.

MORGAN-JONES, G., T.R. NAG RAJ & W.B. KENDRICK (1972) - Icones generum coelomycetum I-V, Waterloo, Ontario.

MORQUER, R., G. VIALA, J. ROUCH, J. FAYRET & G. BERGÉ (1963) - Contribution à l'étude morphogénique du genre Gliocladium. Bull. Soc. mycol. Fr. *79*: 137-241.

MORRALL, R. A. A. (1968) - Two new species of Oidiodendron from boreal forest soils. Can. J. Bot. *46*: 203-206.

MORRIS, E. F. (1963) - The synnematous genera of the Fungi Imperfecti. Western Illin. Univ., Ser. biol. Sci. *3*, 143 pp.

MORTON, F. J. & G. SMITH (1963) - The genera Scopulariopsis, Microascus and Doratomyces. Mycol. Pap. *86*, 96 pp.

MOUCHACCA, J. (1971) - Pseudeurotium desertorum. Revue Mycol. *36*: 123-127.

MÜLLER, E. (1950) - Die schweizerischen Arten der Gattung Leptosphaeria und ihrer Verwandten. Sydowia *4*: 185-319.

MÜLLER, E. (1951) - Die schweizerischen Arten der Gattungen Clathrospora, Pleospora, Pseudoplea und Pyrenophora. Sydowia *5*: 248-310.

MÜLLER, E. (1952) - Die schweizerischen Arten der Gattung Ophiobolus. Ber. schweiz. bot. Ges. *62*: 307-339.

MÜLLER, E. (1957) - Haupt- und Nebenfruchtformen bei Guignardia. Sydowia, Beih. *1*: 210-224.

MÜLLER, E. (1959) - Über die Stellung der Ascomycetengattung Wawelia. Omagiu T. Savulescu p. 515-518.

MÜLLER, E. & J. A. von ARX (1962) - Die Gattungen der didymosporen Pyrenomyceten. Beitr. Krypt.Fl. Schweiz *11*, 2, 922 pp.

MÜLLER, E. & J.A. von ARX (1973) - Pyrenomycetes: Meliolales, Coronophorales, Sphaeriales. In: The Fungi, vol. IV A: 87-132.

MÜLLER, E. & R. W. G. DENNIS (1965) - Fungi venezuelani VIII. Kew Bull. *19*: 357-386.

MÜLLER, E., R. HÜTTER & H. SCHÜEPP (1958) - Über einige bemerkenswerte Discomyceten aus den Alpen. Sydowia *12*: 404-430.

MÜLLER, E. & R. PACHA-AUE (1968) - Untersuchungen an drei Arten von Arachniotus. Nova Hedwigia *15*: 551-558.

MÜLLER, G. (1964, 1965) - Die Gattung Sporotrichum. Eine taxonomische und morphologische Studie der bei Mensch und Tier vorkommenden Spezies. Wiss. Z. Humboldt Univ. Berlin, math.-nat. Reihe, *13*: 611-638, 843-860; *14*: 753-798.

MUNDKUR, B. B. & M. J. THIRUMALACHAR (1952) - Ustilaginales of India. C.M.I., 84 pp.

MUNK, A. (1957) - Danish Pyrenomycetes. Dansk bot. Ark. *17*, 491 pp.

NABEL, K. (1939) - Über die Membran niederer Pilze, besonders von Rhizidiomyces bivelatus. Arch. Mikrobiol. *10*: 515-541.

NAND, K. & B.S. MEHROTRA (1968) - Species of Pilobolus and Pilaira from India. Sydowia *22*: 299-306

NEERGAARD, P. (1945) - Danish Species of Alternaria and Stemphylium. Copenhagen, 559 pp.

NELSON, A. C. & M. P. BACKUS (1968) - Ascocarp development in two homothallic Neurosporas. Mycologia *60*: 16-28.

NELSON, R. R. (1960) - A correlation of interspecific fertility and conidial morphology in species of Helminthosporium exhibiting bipolar germination. Mycologia *52*: 753-761.

NELSON, R. R. (1964) - The perfect stage of Curvularia geniculata. Mycologia *56*: 777-779.

NELSON, R. R. & F. A. HAASIS (1964) - The perfect stage of Curvularia lunata. Mycologia *56*: 316-317.

NEWHALL, A. G. (1946) - More on the name Ansatospora acerina. Phytopathology *36*: 893-896.

NICOT, J. (1948) - Sur quelques Papulospora du sol. Bull. Soc. mycol. Fr. *64*: 209-222.

NICOT, J. (1951) - Revue systématique du genre Cylindrocarpon. Rev. Mycol. *16*: 36-61.

NICOT, J. (1962) - La flore des moisissures d'un Polypore. Rev. Mycol. *27*: 87-92.

NICOT, J. & F. DURAND (1965) - Remarques sur la moisissure fongicole Amblyosporium botrytis. Bull. Soc. mycol. Fr. *81*: 623-649.

NICOT, J. & C. OLIVRY (1961) - Contribution à l'étude du genre Myrothecium Tode. I. Les espèces à spores striées. Rev. gén. Bot. *68*: 673-685.

NICOT, J. & J. ROUCH (1965) - Développement et sporogénèse d'une sphaeropsidale isolée d'un sol de vignoble Toulousain. Ann. Sci. nat., Bot., Sér. 12, *6*: 769-779.

NICOT, J. & V. ZAKARTCHENKO (1966) - Remarques sur la morphologie et la biologie du Cladosporium resinae. Rev. Mycol. *31*: 48-74.

NILSSON, S. (1964) - Freshwater Hyphomycetes. Symb. bot. uppsal. *18*, 2, 130 pp.

NOBLES, M.K. (1948) - Studies in Forest Pathology VI. Can. J. Res., Sect. C, *26*: 281-431.

NOBLES, M. K. (1965) - Identification of cultures of wood-inhabiting Hymenomycetes. Can. J. Bot. *43*: 1097-1139.

NOVAK, R. O. & M. P. BACKUS (1963) - A new species of Mycotypha with a zygosporic stage. Mycologia *55*: 790-798.

NÜESCH, J. (1960) - Beitrag zur Kenntnis der weidenbewohnenden Venturiaceae. Phytopath. Z. *39*: 329-360.

NYLAND, G. (1949) - Studies on some unusual Heterobasidiomycetes from Washington State. Mycologia *41*: 686-701.

NYLAND, G. (1950) - The genus Tilletiopsis. Mycologia *42*: 487-496.

OBRIST, W. (1961) - The genus Ascodesmis. Can. J. Bot. *39*: 943-953.

OMVIK, A. (1955) - Two new species of Chaetomium and one new Humicola species. Mycologia 47: 748-757.

ONDREJ, M. (1971) - Fungi of the genus Fusicladium developing conidia in chains. Česká Mykol. 25: 165-172.

ONIONS, A. H. S. (1968) - Acremoniella velata sp. nov. Trans. Br. mycol. Soc. 51: 151-152.

ONIONS, A. H. S. & G. L. BARRON (1967) - Monophialidic species of Paecilomyces. Mycol. Pap. 107, 25 pp.

ORR, G. F. & H. H. KUEHN (1963) - The genus Ctenomyces. Mycopath. Mycol. appl. 21: 321-333.

ORR, G. F., H. H. KUEHN & O. A. PLUNKETT (1963) - The genus Myxotrichum. Can. J. Bot. 41: 1457-1480.

ORR, G. F., H. H. KUEHN & E. VARSAVSKY (1965) - The genus Amauroascus. Mycopath. Mycol. appl. 25: 100-108.

OWEN, H. (1958) - Physiologic specialization in Rhynchosporium secalis. Trans. Br. mycol. Soc. 41: 99-108.

PADHYE, A.A. & J.W. CARMICHAEL (1971) - The genus Arthroderma. Can. J. Bot. 49: 1525-1540.

PAGE, R. M. (1956) - Studies on the development of asexual reproductive structures in Pilobolus. Mycologia 48: 206-224.

PAPENDORF, M. C. (1967) - Two new genera of soil fungi from South Africa. Trans. Br. mycol. Soc. 50: 69-75.

PAPENDORF, M. C. (1969) - New South African soil fungi. Trans. Br. mycol. Soc. 52: 483-489.

PAPENDORF, M. C. & J. A. von ARX (1966) - Herpotrichia striatispora, a new ascomycete from South Africa. Nova Hedwigia 12: 395-397.

PAPENDORF, M. C. & H. P. UPADHYAY (1969) - Botryoderma lateritium and B. rostratum gen. et spp. nov. from soil in South Africa and Brazil. Trans. Br. mycol. Soc. 52: 257-265.

PARBERY, D.G. (1969) - Amorphotheca resinae, the perfect state of Cladosporium resinae. Aust. J. Bot. 17: 331-357.

PARK, J. Y. & R. SPRAGUE (1953) - Studies on some Selenophoma species on Gramineae. Mycologia 45: 260-275.

PATERSON, R.A. (1963) - Observations on two species of Rhizophydium from northern Michigan. Trans. Br. mycol. Soc. 46: 530-536.

PEEK, CH. H. & W. G. SOLHEIM (1958) - The hyphomycetous genera of H. W. Harkness and the ascomycetous genus Cleistosoma. Mycologia 50: 844-861.

PETCH, T. (1938) - British Hypocreales. Trans. Br. mycol. Soc. 21: 243-305.

PETCH, T. (1945) - Stilbum tomentosum. Trans. Br. mycol. Soc. 28: 101-109.

PETERSEN, R. H. (1962) - Spore formation in Tricellula and Volucrispora. Bull. Torrey bot. Club 89: 287-293.

PETERSEN, R. H. (1962) - Aquatic Hyphomycetes from North America. I. Aleuriosporae (Part 1) and key to the genera. Mycologia 54: 117-151.

PETERSEN, R. H. (1963) - Aquatic Hyphomycetes from North America. II and III. Mycologia 55: 18-29, 570-581.

PETRAK, F. (1925) - Mykologische Notizen VII. Ann. mycol. 23: 1-143.

PETRAK, F. (1941) - Mykologische Notizen XIV. Ann. mycol. 39: 251-349.

PETRAK, F. (1947) - Über die Gattung Albertiniella. Sydowia 1: 83-85.

PETRAK, F. (1953) - Über die Gattung Septogloeum. Sydowia 7: 313-315.

PETRAK, F. (1956) - Über die Gattungen Dothichiza und Chondroplea. Sydowia 10: 201-235.

PETRAK, F. (1964) - Über die Gattungen Chaetomella, Amerosporium und Volutellospora. Sydowia *18*: 373-379 (1965).

PETRAK, F. (1966) - Über die Gattungen Petrakia und Echinosporium. Sydowia *20*: 186-189 (1968).

PETRAK, F. & H. SYDOW (1927) - Die Gattungen der Pyrenomyzeten, Sphaeropsideen und Melanconieen. 1. Teil: Die phaeosporen Sphaeropsideen und die Gattung Macrophoma. In Fedde Rep. Spec. nov. Reg. veg., Beih. *42*: 551 pp.

PIDOPLICHKO, N.M. & A.A. MILKO (1971) - Atlas mukoralnykh Gribov. Kiev (in Russian).

PIROZYNSKI, K. A. (1963) - Beltrania and related genera. Mycol. Pap. *90*, 37 pp.

PIROZYNSKI, K. A. (1969) - Reassessment of the genus Amblyosporium. Can. J. Bot. *47*: 325-334.

PITT, J. I. (1966) - Two new species of Chrysosporium. Trans. Br. mycol. Soc. *49*: 467-470.

PITT, J. I. & M. W. MILLER (1968) - Sporulation in Candida pulcherrima, Candida reukaufii and Chlamydozyma species: their relationships with Metschnikowia. Mycologia *60*: 663-685.

PLAATS—NITERINK, A. J. van der (1968) - The occurrence of Pythium in the Netherlands. I. Heterothallic species. Acta bot. neerl. *17*: 320-329.

PRESTON, N. C. (1943, 1948, 1961) - Observations on the genus Myrothecium. Trans. Br. mycol. Soc. *26*: 158-168; *31*: 271-276; *44*: 31-41.

PRIDHAM, T. G. & K. B. RAPER (1950) - Ashbya gossypii - its significance in nature and in the laboratory. Mycologia *42*: 603-623.

PROWSE, G. A. (1954) - Aphanomyces daphniae sp. nov., parasitic on Daphnia hyalina. Trans. Br. mycol. Soc. *37*: 22-28.

PUGH, G. J. F., J. P. BLAKEMAN & G. MORGAN—JONES (1964) - Thermomyces verrucosus and T. lanuginosus. Trans. Br. mycol. Soc. *47*: 115-121.

PUGH, G. J. F. & J. NICOT (1964) - Studies on fungi in coastal soils. V. Dendryphiella salina. Trans. Br. mycol. Soc. *47*: 263-267.

PURDY, L. H. (1955) - A broader concept of the species Sclerotinia sclerotiorum based on variability. Phytopathology *45*: 421-427.

RAI, J. N. & J. P. TEWARI (1963) - A new Acremoniella from Indian soils. Can. J. Bot. *41*: 331-334.

RAI, J.N. & K. WADHWANI (1970) - Anixiella indica. Studies in ascocarp development and ascus cytology. J. gen. appl. Microbiol. *16*: 251-258.

RAO, P. N. (1966) - A new species of Dichotomophthora on Portulaca oleracea from Hyderabad, India. Mycopath. Mycol. appl. *28*: 137-140.

RAO, P. R. (1963) - A new species of Myrothecium from soil. Antonie van Leeuwenhoek *29*: 180-182.

RAO, P. R. & D. RAO (1964) - The genus Periconia in India. Mycopath. Mycol. appl. *22*: 285-310.

RAO, V. G. (1965) - The fungus genus Alternaria in Bombay-Maharashtra I. and II. Sydowia *18*: 44-64, 65-85.

RAO, V. G. (1969) - The genus Alternaria from India. Nova Hedwigia *17*: 219-258.

RAPER, K. B. (1935) - Dictyostelium discoideum, a new species of slime mold from decaying forest leaves. J. agric. Res. *50*: 135-147.

RAPER, K. B. (1951) - Isolation, cultivation and conservation of simple slime molds. Quart. Rev. Biol. *25*: 169-190.

RAPER, K. B. & D. I. FENNELL (1965) - The genus Aspergillus. Baltimore, 686 pp.

RAPER, K. B. & C. THOM (1949) - A manual of the Penicillia. Baltimore, 875 pp.

RAPILLY, F. (1966) - Limites proposées au genre Helminthosporium, relations avec les genres voisins. Bull. Soc. mycol. Fr. *82*: 221-240.

REBELL, G., D. TAPLIN & H. BLANK (1964) - Dermatophytes, their recognition and identification. Miami, 57 pp.

REISINGER, O. (1968) - Remarques sur les genres Dendryphiella et Dendryphion. Bull. Soc. mycol. Fr. 84: 27-51.

REMY, E. (1950) - Über niedere Bodenphycomyceten. Arch. Mikrobiol. 14: 212-239.

REUSSER, F. A. (1964) - Über einige Arten der Gattung Guignardia. Phytopath. Z. 51: 205-240.

RIFAI, M. A. (1964) - Stachybotrys bambusicola sp. nov. Trans. Br. mycol. Soc. 47: 269-272.

RIFAI, M. A. (1965) - On Sporidesmium trigonellum. Persoonia 3: 407-411.

RIFAI, M. A. (1968) - The australasian Pezizales in the herbarium of the Royal botanic gardens Kew. Verh. K. ned. Akad. Wet., afd. Natuurk., 2de reeks, 57 (3), 295 pp.

RIFAI, M. A. (1969) - A revision of the genus Trichoderma. Mycol. Pap. 116, 56 pp.

RIFAI, M. A. & R. C. COOKE (1966) - Studies on some didymosporous genera of nematode-trapping Hyphomycetes. Trans. Br. mycol. Soc. 49: 147-168.

RIMPAU, R. H. (1961) - Untersuchungen über die Gattung Drepanopeziza. Phytopath. Z. 43: 257-306.

ROBERTS, C. (1960) - The life cycle of Kluyveromyces africanus. C. r. Trav. Lab. Carlsberg 31: 325-341.

ROBERTS, C. & J. P. van der WALT (1959) - The life cycle of Kluyveromyces polysporus. C. r. Trav. Lab. Carlsberg 31: 129-148.

ROBINSON, W. (1926) - The conditions of growth and development of Pyronema confluens. Ann. Bot. 40: 245-272.

ROBINSON-JEFFREY, R.C. & R.W. DAVIDSON (1968) - Three new Europhium species with Verticicladiella imperfect states on blue stained pine. Can. J. Bot. 46: 1523-1527.

RODRIGUES de MIRANDA, L. (1972) - Filobasidium capsuligenum. Antonie van Leeuwenhoek 38: 91-99.

ROSSY—VANDERRAMA, C. (1956) - Some water molds from Puerto Rico. J. Elisha Mitchell scient. Soc. 72: 129-137.

ROY, R. Y., R. S. DWIVEDI & R. R. MISHRA (1962) - Two new species of Scolecobasidium from soil. Lloydia 25: 164-166.

ROY, R. Y. & S. GUJARATI (1965) - A new species of Dactylaria from soil. Lloydia 28: 53-54.

RUPPRECHT, H. (1959) - Beiträge zur Kenntnis der Fungi imperfecti III. Sydowia 13: 10-22.

SAËZ, H. E. (1957) - Le Geotrichum candidum. Caractéristiques morpho-biologiques; fréquence chez l'homme. Bull. Soc. mycol. Fr. 73: 343-353.

SAHTIYANCI, S. (1962) - Studien über einige wurzelparasitäre Olpidiaceen. Arch. Mikrobiol. 41: 187-228.

SAKSENA, H. K. & O. VAARTAJA (1960) - Descriptions of new species of Rhizoctonia. Can. J. Bot. 38: 931-943.

SAKSENA, H. K. & O. VAARTAJA (1961) - Taxonomy, morphology and pathogenicity of Rhizoctonia species from forest nurseries. Can. J. Bot. 39: 627-674.

SAMSON, R. A. (1969) - Revision of the genus Cunninghamella (Fungi, Mucorales). Proc. K. ned. Akad. Wet., C, 72: 322-335.

SAMSON, R.A. (1972) - Notes on Pseudogymnoascus, Gymnoascus and related genera. Acta bot. neerl. 21: 517-527.

SAMSON, R.A. & H.A. van der AA (1973) - Het ascomycetengeslacht Onygena. Coolia 16: 30-35.

SAMSON, R.A. & T. MAHMOOD (1970) - The genus Acrophialophora. Acta bot. neerl. 19: 804-808.

SCHEINPFLUG, H. (1958) - Untersuchungen über die Gattung Didymosphaeria und einige verwandte Gattungen. Ber. schweiz. bot. Ges. *68*: 325-385.

SCHIPPER, M. A. A. (1967) - Mucor strictus, a psychrophilic fungus, and Mucor falcatus sp. n. Antonie van Leeuwenhoek *33*: 189-195.

SCHIPPER, M. A. A. (1967) - On Mucor fuscus and its synonyms. Antonie van Leeuwenhoek *33*: 473-476.

SCHIPPER, M. A. A. (1969) - Zygosporic stages in heterothallic Mucor. Antonie van Leeuwenhoek *35*: 189-208.

SCHIPPER, M. A. A. (1973) - A study on variability in Mucor hiemalis and related species. Stud. Mycol. *4*, 40 pp.

SCHIPPER, M. A. A. & V. HINTIKKA (1969) - Zygorhynchus psychrophilus sp. n. Antonie van Leeuwenhoek *35*: 29-32.

SCHIPPERS–LAMMERTSE, A. F. & C. HEYTING (1962) - Physiological properties, conjugation and taxonomy of Cephaloascus fragrans. Antonie van Leeuwenhoek *28*: 5-16.

SCHLÄPFER-BERNHARD, E. (1969) - Beitrag zur Kenntnis der Discomycetengattungen Godronia, Ascocalyx, Neogodronia und Encoeliopsis. Sydowia *22*: 1-56.

SCHLÖSSER, U. G. (1970) - Mastigosporium kitzebergense spec. nov., ein parasitischer Pilz auf Phleum pratense. Phytopath. Z. *67*: 248-258.

SCHNEIDER, R. (1954) - Plicaria fulva n. sp;, ein bisher nicht bekannter Gewächshausbewohner. Zentbl. Bakt. Parasitkde, Abt. II, *108*: 147-153.

SCHNEIDER, R. & W. GERLACH (1966) - Pyrenochaeta lycopersici nov. spec., der Erreger der Korkwurzelkrankheit der Tomate. Phytopath. Z. *56*: 117-122.

SCHOL–SCHWARZ, M. B. (1959) - The genus Epicoccum. Trans. Br. mycol. Soc. *42*: 149-173.

SCHOL–SCHWARZ, M. B. (1968) - Rhinocladiella, its synonym Fonsecaea and its relation to Phialophora. Antonie van Leeuwenhoek *34*: 119-152.

SCHOL–SCHWARZ, M. B. (1970) - Revision of the genus Phialophora. Persoonia *6*: 59-94.

SCHRANTZ, J. P. (1960) - Recherches sur les Pyrénomycètes de l'ordre de Diatrypales. Bull. Soc. mycol. Fr. *76*: 305-407.

SCOTT, De B. (1968) - The genus Eupenicillium. S. Afr. C.S.I.R. Pretoria, 150 pp.

SCOTT, De B. & A. C. STOLK (1967) - Studies on the genus Eupenicillium. II. Perfect states of some Penicillia. Antonie van Leeuwenhoek *33*: 297-314.

SCOTT, W.W. (1961) - A monograph of the genus Aphanomyces. Tech. Bull. Va agric. Exp. Stn *151*: 1-95.

SEELER, E. V. (1943) - Several fungicolous fungi. Farlowia *1*: 119-133.

SEEMÜLLER, E. (1968) - Untersuchungen über die morphologische und biologische Differenzierung in der Fusarium-Sektion Sporotrichiella. Mitt. biol. Bundesanst. Ld- u. Forstw. *127*, 93 pp.

SERVAZZI, O. (1939) - Contributi alla patologia dei pioppi VI. Boll. Lab. Sperim. R. Oss. Fitopat. Torino *15*: 49-152.

SETH, H. K. (1967) - Studies on the genus Chaetomium. I. Heterothallism. Mycologia *59*: 580-584.

SETH, H. K. (1968) - Chaetomidium trichorobustum sp. nov. from Germany. Nova Hedwigia *16*: 429-432.

SETH, H. K. (1968) - The fungus genus Kernia with the description of a new species. Acta bot. neerl. *17*: 478-482.

SETH, H.K. (1970) - The genus Lophotrichus. Nova Hedwigia *19*: 591-599 (publ. 1971).

SETH, H.K. (1970) - A monograph of the genus Chaetomium. Beih. Nova Hedwigia *37*, 133 pp. (publ. 1972).

SEYMOUR, R. L. (1970) - The genus Saprolegnia. Nova Hedwigia *19*: 1-160.

SHEAR, C. L. (1922) - Life history of an undescribed ascomycete isolated from granular mycetoma of man. Mycologia *14*: 239-243.

SHIGO, A. L. (1960) - Mycoparasitism of Gonatobotryum fuscum and Piptocephalis xenophila. Trans. N.Y. Acad. Sci. *22*: 365-372.

SHOEMAKER, R. A. (1955) - Biology, cytology and taxonomy of Cochliobolus sativus. Can. J. Bot. *33*: 562-576.

SHOEMAKER, R. A. (1959) - Nomenclature of Drechslera and Bipolaris, grass-parasites segregated from Helminthosporium. Can. J. Bot. *37*: 879-887.

SHOEMAKER, R. A. (1961) - Pyrenophora phaeocomes. Can. J. Bot. *39*: 901-908.

SHOEMAKER, R. A. (1962) - Drechslera. Can. J. Bot. *40*: 809-836.

SHOEMAKER, R. A. (1964) - Seimatosporium (= Cryptostictis) parasites of Rosa, Vitis and Cornus. Can. J. Bot. *42*: 411-421.

SHOEMAKER, R. A. (1964) - Generic correlations and concepts: Clathridium (= Griphosphaeria) and Seimatosporium (= Sporocadus). Can. J. Bot. *42*: 403-407.

SHOEMAKER, R. A. (1966) - A pleomorphic parasite of cereal seeds, Pyrenophora seminiperda. Can. J. Bot. *44*: 1451-1456.

SHOEMAKER, R. A. (1968) - Type studies of Pleospora calvescens, Pleospora papaveracea and some allied species. Can. J. Bot. *46*: 1143-1150.

SHOEMAKER, R. A. & E. MÜLLER (1963) - Generic correlations and concepts: Broomella and Pestalotia. Can. J. Bot. *41*: 1235-1243.

SHOEMAKER, R. A. & E. MÜLLER (1965) - Types of the pyrenomycete genera Hymenopleella and Lepteutypa. Can. J. Bot. *43*: 1457-1460.

SIMMONS, E. G. (1952) Culture studies in the genera Pleospora, Clathrospora and Leptosphaeria. Mycologia *44*: 330-365.

SIMMONS, E. G. (1954) - Culture studies in the genera Pleospora, Clathrospora and Leptosphaeria II. Mycologia *46*: 184-200.

SIMMONS, E. G. (1967) - Typification of Alternaria, Stemphylium and Ulocladium. Mycologia *59*: 67-92.

SIMMONS, E. G. (1969) - Perfect states of Stemphylium. Mycologia *61*: 1-26.

SIMMONDS, J. H. (1965) - A study of the species of Colletotrichum causing ripe fruit rots in Queensland. Queensl. J. agr. anim. Sci. *22*: 437-459.

SINHA, S. (1940) - A wet rot of Colocasia antiquorum due to secondary infection by Choanephora cucurbitarum and Choaneophora trispora (= Blakeslea trispora). Proc. Indian Acad. Sci. *11*: 167-176.

SIVANESAN, A. (1971) - The genus Herpotrichia. Mycol. Pap. *127*, 37 pp.

SKOLKO, A. J. & J. W. GROVES (1948) - Notes on seed-borne fungi. V. Chaetomium species with dichotomously branched hairs. Can. J. Res., Sect. C. *26*: 269-280.

SKOLKO, A. J. & J. W. GROVES (1953) - Notes on seed-borne fungi. VII. Chaetomium. Can. J. Bot. *31*: 779-809.

SKOU, J. P. (1968) - Studies on the take-all fungus, Gaeumannomyces graminis I. Yb. R. vet. agric. Coll. Copenhagen 1968, 109-116.

SKOU, J. P. (1972) - Ascosphaerales. Friesia *10*,1: 1-24.

SMERLIS, E. (1962) - Taxonomy and morphology of Potebniamyces balsamicola associated with a twig and branch blight of Balsam Fir in Quebec. Can. J. Bot. *40*: 351-359.

SMITH, G. (1962) - Some new and interesting species of microfungi. III. Trans. Br. mycol. Soc. *45*: 387-394.

SMITH, G. (1965) - Three new species of Penicillium. Trans. Br. mycol. Soc. *48*: 273-277.

SOWELL, G. & R. P. KORF (1962) - An emendation of the genus Itersonilia based on studies of morphology and pathogenicity. Mycologia 52: 934-945.

SPARROW, F. K. (1960) - Aquatic Phycomycetes. Ann Arbor, 1187 pp.

SPARROW, F. K. & M. E. BARR (1955) - Additions to the Phycomycete flora of the Douglas lake region. I. New taxa and records. Mycologia 47: 546-556.

SPENCER, J. F. T., H. J. PHAFF & N. R. GARDNER (1964) - Metschnikowia kamienskii sp. n., a yeast associated with brine shrimp. J. Bact. 88: 758-762.

SPRAGUE, R. (1950) - Diseases of cereals and grasses in North America. New York, 538 pp.

SPRAGUE, R. (1957) - Some leafspot fungi on Western gramineae XI. Mycologia 49: 837-853.

SRINIVASAN, M. C. & M. J. THIRUMALACHAR (1963) - Studies on species of Conidiobolus from India II. Sydowia 16: 60-66.

SRINIVASAN, M. C. & M. J. THIRUMALACHAR (1963) - Studies on species of Conidiobolus from India III. Mycologia 54: 685-693.

SRINIVASAN, M. C. & M. J. THIRUMALACHAR (1967) - Evaluation of taxonomic characters in the genus Conidiobolus, with key to species. Mycologia 59: 698-713.

STANDEN, J. H. (1943) - Variability of Nigrospora on maize. Iowa St. Coll. J. Sci. 17: 263-275.

STEYAERT, R. H. (1949) - Contribution à l'étude monographique de Pestalotia et Monochaetia (Truncatella gen. nov. et Pestalotiopsis gen. nov.). Bull. Jard. bot. Brux. 19: 285-354.

STOLK, A. C. (1955) - Emericellopsis minima and Westerdijkella ornata. Trans. Br. mycol. Soc. 38: 419-424.

STOLK, A. C. (1963) - The genus Chaetomella. Trans. Br. mycol. Soc. 46: 409-425.

STOLK, A. C. (1965) - Thermophilic species of Talaromyces and Thermoascus. Antonie van Leeuwenhoek 31: 262-276.

STOLK, A. C. (1969) - Four new species of Penicillium. Antonie van Leeuwenhoek 35: 261-274.

STOLK, A. C. & G. L. HENNEBERT (1968) - New species of Thysanophora and Custingophora gen. nov. Persoonia 5: 189-199.

STOLK, A. C. & D. B. SCOTT (1967) - Studies on the genus Eupenicillium Ludwig. I. Persoonia 4: 391-405.

STOLK, A.C. & R.A. SAMSON (1971) - Studies in Talaromyces and related genera I. Hamigera and Byssochlamys. Persoonia 6: 341-357.

STOLK, A.C. & R.A. SAMSON (1972) - The genus Talaromyces. Studies on Talaromyces and related genera II. Stud. Mycol. 2: 64 pp.

SUBRAMANIAN, C. V. (1953) - Fungi Imperfecti from Madras. IV. Proc. Indian Acad. Sci. 37: 96-105.

SUBRAMANIAN, C. V. (1954) - Fungi Imperfecti from Madras. VI. J. Indian bot. Soc. 33: 36-42.

SUBRAMANIAN, C. V. (1955) - Some species of Periconia from India. J. Indian bot. Soc. 34: 339-361.

SUBRAMANIAN, C. V. (1956) - Hyphomycetes I and II. J. Indian bot. Soc. 35: 53-91 and 446-494.

SUBRAMANIAN, C. V. (1957) - Two new species of Petrakia. Sydowia, Beih. 1: 14-16.

SUBRAMANIAN, C. V. (1958) - Hyphomycetes V and VI. J. Indian bot. Soc. 37: 47-64 and 401-407.

SUBRAMANIAN, C. V. (1962) - The classification of the Hyphomycetes. Bull. bot. Surv. India 4: 249-259.

SUBRAMANIAN, C. V. (1963) - Dactylella, Monacrosporium and Dactylina. J. Indian bot. Soc. 42: 291-300.

SUBRAMANIAN, C.V. (1971) - Hyphomycetes. New Dehli, 930 pp.

SUBRAMANIAN, C.V. (1972) - The perfect states of Aspergillus. Curr. Sci. *41*: 755-761.

SUBRAMANIAN, C. V. & B. L. JAIN (1966) - A revision of some graminicolous Helmintho-sporia. Curr. Sci. *35*: 352-355.

SUBRAMANIAN, C. V. & B. C. LODHA (1964) - Four new coprophilous Hyphomycetes. Antonie van Leeuwenhoek *30*: 317-330.

SUBRAMANIAN, C. V. & B. C. LODHA (1964) - Two interesting Hyphomycetes. Can. J. Bot. *42*: 1057-1063.

SUGIYAMA, J. (1968) - The taxonomic status of the genus Chalaropsis (Hyphomycetes). J. Fac. Sci. Univ. Tokyo 3, *10*: 29-48.

SUGIYAMA, J. (1969) - Studies on Himalayan yeasts and moulds II. Trans. mycol. Soc. Japan *9*: 117-124.

SUGIYAMA, J., Y. KAWASAKI & H. KURATA (1968) - Wardomyces simplex, a new hyphomycete from milled rice. Bot. Mag. Tokyo *81*: 243-250.

SUNDSTRÖM, K. R. (1964) - Studies on the physiology, morphology and serology of Exobasidium. Symb. bot. uppsal. *18* (3), 89 pp.

SUTTON, B. C. (1962) - Colletotrichum dematium and C. trichellum. Trans. Br. mycol. Soc. *45*: 222-232.

SUTTON, B. C. (1963) - Coelomycetes II. Mycol. Pap. *88*, 50 pp.

SUTTON, B. C. (1964) - Melanconium Link ex Fr. Persoonia *3*: 193-198.

SUTTON, B. C. (1964) - Coelomycetes III. Mycol. Pap. *97*, 42 pp.

SUTTON, B. C. (1965) - Typification of Dendrophoma and a reassessment of D. obscurans. Trans. Br. mycol. Soc. *48*: 611-616.

SUTTON, B.C. (1968) - Kellermania and its generic segregates. Can. J. Bot. *46*: 181-196.

SUTTON, B.C. (1969) - Type studies of Coniella, Anthasthoopa and Cyclodomella. Can. J. Bot. *47*: 603-608.

SUTTON, B.C., K.A. PIROZYNSKI & F.C. DEIGHTON (1972) - Microdochium. Can. J. Bot. *50*: 1849-1907.

SWART, H. J. (1959) - A comparative study of the genera Gonytrichum and Bisporomyces. Antonie van Leeuwenhoek *25*: 439-444.

SWART, H. J. (1964) - A study of the production of coremia in three species of the genus Trichurus. Antonie van Leeuwenhoek *30*: 257-260.

SWART, H. J. (1964) - A new fungus record for South Africa. S. Afr. J. Sci. *60*: 287.

SWART, H. J. (1965) - Conidial formation in Haplographium fuligineum. Trans. Br. mycol. Soc. *48*: 459-461.

SWART, H.J. (1973) - The fungus causing cypress canker. Trans. Br. mycol. Soc. *61*: 71-82.

SYDOW, H. (1926) - Fungi in itinere costaricensi collecti. Ann. mycol. *24*: 283-426.

SYDOW, H. & F. PETRAK (1928) - Micromycetes philippinenses, Series prima. Ann. mycol. *26*: 414-446.

TEHON, L. R. (1948) - Notes on the parasitic fungi of Illinois. Mycologia *40*: 314-327.

TERRIER, CH. A. (1942) - Essai sur la systématique des Phacidiaceae. Berne, 99 pp.

THOMSON, J. R. (1936) - Cylindrosporium concentricum. Trans. Br. mycol. Soc. *20*: 123-132.

TIMONIN, M. I. (1961) - New species of the genus Pseudobotrytis. Ceiba *9*: 27-29.

TUBAKI, K. (1952) - Studies on the Sporobolomycetaceae in Japan. I. On Tilletiopsis. Nagaoa *1*: 26-31.

TUBAKI, K. (1955) - Studies on Japanese Hyphomycetes. II. Fungicolous group. Nagaoa *5*: 11-40.

TUBAKI, K. (1957) - Biological and cultural studies of three species of Protomyces. Mycologia *49*: 44-54.

TUBAKI, K. (1958) - Studies on the Japanese Hyphomycetes V. Leaf and stem group with a discussion of the classification of Hyphomycetes and their perfect stages. J. Hattori bot. Lab. *20*: 142-244.

TUBAKI, K. (1963) - Taxonomic study of Hyphomycetes. A. Rep. Inst. Ferment., Osaka (1961-62) *1*: 25-54.

TUBAKI, K. (1963) - Notes on the Japanese Hyphomycetes. I. Chloridium, Clonostachys, Isthmospora, Pseudobotrytis, Stachybotrys and Stephanoma. Trans. mycol. Soc. Japan *4*: 83-90.

TUBAKI, K. (1967) - An undescribed species of Heleococcum from Japan. Trans. mycol. Soc. Japan *8*: 5-10.

TUBAKI, K. & M. SONEDA (1959) - Cultural and taxonomical studies on Prototheca. Nagaoa *6*: 25-34.

TUCKER, C. M. (1931) - The taxonomy of the genus Phytophthora. Bull. Mo. agric. Exp. Stn *153*, 208 pp.

TULLOCH, M. (1972) - The genus Myrothecium. Mycol. Pap. *130*, 42 pp.

TURNER, M. (1963) - Studies in the genus Mortierella - I. M. isabellina and related species. Trans. Br. mycol. Soc. *46*: 262-272.

UDAGAWA, S. (1960) - A taxonomic study of the Japanese species of Chaetomium. J. gen. appl. Microbiol. *6*: 223-251.

UDAGAWA, S. (1962) - Microascus species new to the Mycoflora of Japan. l.c. *8*: 39-51.

UDAGAWA, S. (1962) - Carpenteles brefeldianum and Myxotrichum chartarum, two japanese records. Trans. mycol. Soc. Japan *4*: 11-13.

UDAGAWA, S. (1963) - Notes on some Japanese Ascomycetes. Trans. mycol. Soc. Japan *4*: 94-102.

UDAGAWA, S. (1963) - Neocosmospora in Japan. Trans. mycol. Soc. Japan *4*: 121-125.

UDAGAWA, S. (1963) - Microascaceae in Japan. J. gen. appl. Microbiol. *9*: 137-148.

UDAGAWA, S. (1965) - Notes on some Japanese Ascomycetes II. Trans. mycol. Soc. Japan *6*: 78-90.

UDAGAWA, S. (1966) - Notes on some Japanese Ascomycetes III. Trans. mycol. Soc. Japan *7*: 91-97.

UDAGAWA, S. & R. F. CAIN (1969) - Notes on the genus Microthecium. Can. J. Bot. *47*: 1915-1933.

UDAGAWA, S. & R. F. CAIN (1969) - Some new or noteworthy species of the genus Chaetomium. Can. J. Bot. *47*: 1939-1951.

UDAGAWA, S. & K. FURUYA (1972) - Zopfiella pilifera, a new cleistoascomycete from Japanese soil. Trans. mycol. Soc. Japan *13*: 255-259.

UDAGAWA, S. & Y. HORIE (1972) - Diplogelasinospora and its conidial state. J. Jap. Bot. *47*: 297-306.

UDAGAWA, S. & Y. KAWASAKI (1968) - Notes on some Japanese Ascomycetes VI. Trans. mycol. Soc. Japan *8*: 115-121.

UDAGAWA, S. & M. TAKADA (1967) - Notes on some Japanese Ascomycetes IV and V. Trans. mycol. Soc. Japan *8*: 43-53.

UDAGAWA, S. & M. TAKADA (1973) - The rediscovery of Aphanoascus cinnabarinus. J. Jap. Bot. *48*: 21-26.

UDEN, N. van & R. CASTELO—BRANCO (1961) - Metschnikowiella zobellii sp. nov. and M. krissii sp. nov., two yeasts from the Pacific Ocean pathogenic for Daphnia magna. J. gen. Microbiol. *26*: 141-148.

UEBELMESSER, E. R. (1956) - Über einige neue Chytridineen aus Erdboden. Arch. Mikrobiol. *25*: 307-324.

UPADHYAY, H. P. (1966) - Soil fungi from North-East Brazil. II. Mycopath. Mycol. appl. *30*: 276-286.

UPADHYAY, H. P. (1970) - Soil fungi from North-east and North Brazil - VIII. Persoonia *6*: 111-117.

UPADHYAY, H.P. (1973) - Helicostylum and Thamnostylum. Mycologia *65*: 733-751.

VANBREUSEGHEM, R. (1952) - Intérêt théorique et pratique d'un nouveau dermatophyte isolé du sol: Keratinomyces ajelloi gen. nov., sp. nov. Bull. Acad. Belg. Clin. Sci. *38*: 1068-1077.

VASSILJEVSKI, N. I. & B. P. KARAKULIN (1950) - Fungi Imperfecti Parasitici, pars II. Melanconiales. Leningrad, 680 pp.

VEEN, K. H. (1968) - Recherches sur la maladie, due à Metarrhizium anisopliae chez le criquet pèlerin. Meded. Landb.Hogesch. Wageningen (5), 77 pp.

VERONA, O. & G. MAZZUCCHETTI (1968) - I generi Stachybotrys e Memnoniella. Roma, 111 pp.

VRIES, G. A. de (1952) - Contribution to the knowledge of the genus Cladosporium. Baarn, 121 pp.

VRIES, G. A. de (1962) - Cyphellophora laciniata nov. gen., nov. sp. and Dactylium fusarioides Fragoso et Ciferri. Mycopath. Mycol. appl. *16*: 47-54.

VRIES, G. A. de (1964) - Keratinophilic fungi. Ann. Soc. belge Méd. trop. *44*: 795-802.

VRIES, G. A. de (1969) - Das Problem Aphanoascus oder Anixiopsis. Mykosen *12*: 111-122.

WAKEFIELD, E. M. & G. R. BISBY (1941) - List of Hyphomycetes recorded for Britain. Trans. Br. mycol. Soc. *25*: 49-126.

WALKER, J. (1972) - Type studies on Gaeumannomyces graminis and related fungi. Trans. Br. mycol. Soc. *58*: 427-457.

WALKER, L. B. (1931) - Studies on Ascoidea rubescens I. History and development. Mycologia *23*: 51-76.

WALT, J. P. van der (1956) - Kluyveromyces, a new yeast genus of the Endomycetales. Antonie van Leeuwenhoek *22*: 265-272.

WALT, J. P. van der (1959) - Endomyces reessii. Antonie van Leeuwenhoek *25*: 458-464.

WALT, J. P. van der (1965) - The emendation of the genus Kluyveromyces. Antonie van Leeuwenhoek *31*: 341-348.

WALT, J. P. van der (1969) - The genus Syringospora. Antonie van Leeuwenhoek *35*, supplement A1.

WALT, J.P. van der & D.B. SCOTT (1971) - The yeast genus Saccharomycopsis. Mycopath. Mycol. appl. *43*: 279-288.

WANG, C. J. K. (1965) - Fungi of pulp and paper in New York. Tech. Publ. Syracuse St. Univ. Coll. Forestry *87*.

WANG, C. J. K. & G. E. BAKER (1967) - Zygosporium masonii and Z. echinosporum from Hawaii. Can. J. Bot. *45*: 1945-1952.

WARCUP, J. H. & P. H. B. TALBOT (1962) - Ecology and identity of mycelia isolated from soil. Trans. Br. mycol. Soc. *45*: 495-518.

WARCUP, J. H. & P. H. B. TALBOT (1963) - Ecology and identity of mycelia isolated from soil. II. Trans. Br. mycol. Soc. *46*: 465-472.

WARCUP, J. H. & P. H. B. TALBOT (1966) - Perfect states of some Rhizoctonias. Trans. Br. mycol. Soc. *49*: 427-435.

WARCUP, J. H. & P. H. B. TALBOT (1967) - Perfect states of Rhizoctonias associated with orchids. New Phytol. *66*: 631-641.

WATERHOUSE, G. M. (1956) - The genus Phytophthora. Diagnoses and figures from the original papers. Misc. Publ. C.M.I., Kew, *12*, 120 pp.

WATERHOUSE, G. M. (1963) - Key to the species of Phytophthora. Mycol. Pap. 92, 22 pp.

WATERHOUSE, G. M. (1967) - Key to Pythium Pringsheim. Mycol. Pap. 109, 15 pp.

WATERHOUSE, G. M. (1968) - The genus Pythium. Diagnoses (or descriptions) and figures from the original papers. Mycol. Pap. 110, 71 pp., 50 plates.

WATSON, P. (1955) - Calcarisporium arbuscula living as an endophyte in apparently healthy sporophores of Russula and Lactarius. Trans. Br. mycol. Soc. 38: 409-414.

WEBSTER, J. (1961) - The Mollisia perfect state of Anguillospora crassa. Trans. Br. mycol. Soc. 44: 559-564.

WEBSTER, J. (1964) - Culture studies on Hypocrea and Trichoderma I. Trans. Br. mycol. Soc. 47: 75-96.

WEBSTER, J. (1965) - The perfect state of Pyricularia aquatica. Trans. Br. mycol. Soc. 48: 449-452.

WEBSTER, J., M. A. RIFAI & M. SAMY EL—ABYAD (1964) - Culture observations on some Discomycetes from burnt ground. Trans. Br. mycol. Soc. 47: 445-454.

WEHMEYER, L. E. (1933) - The genus Diaporthe. Univ. Mich. Stud. 9, 349 pp.

WEHMEYER, L. E. (1941) - A revision of Melanconis, Pseudovalsa, Prosthecium and Titania. Univ. Mich. Stud. 14, 161 pp.

WEHMEYER, L. E. (1961) - A world monograph of the genus Pleospora and its segregates. Ann Arbor, 451 pp.

WERESUB, L.K. & P.M. LE CLAIR (1971) - On Papulaspora and bulbilliferous basidiomycetes Burgoa and Minimedusa. Can. J. Bot. 49: 2203-2213.

WHETZEL, H. H. (1945) - A synopsis of the genera and species of the Sclerotiniaceae, a family of stromatic inoperculate Discomycetes. Mycologia 37: 648-714.

WHITE, W. L. (1941) - A monograph of the genus Rutstroemia. Lloydia 4: 153-240.

WHITE, W. L. & M. H. DOWNING (1953) - Humicola grisea, a soil inhabiting cellulolytic Hyphomycete. Mycologia 45: 951-963.

WHITTAKER, R. H. (1969) - New concepts of kingdoms of organisms. Science 163: 150-160.

WICKERHAM, L. J. (1951) - Taxonomy of yeasts. Tech. Bull. U.S. Dep. Agric. Wash. 1029, 56 pp.

WILEY, B.J. & E.G. SIMMONS (1971) - Gliocephalotrichum, new combinations and a new species. Mycologia 63: 575-585.

WILSON, C. M. (1961) - A cytological study of Ascocybe. Can. J. Bot. 39: 1605-1607.

WILTSHIRE, S. P. (1938) - The original and modern concepts of Stemphylium. Trans. Br mycol. Soc. 21: 211-239.

WINDISCH, S. (1940) - Entwicklungsgeschichtliche Untersuchungen an Torulopsis pulcherrina und Candida tropicalis. Arch. Mikrobiol. 11: 368-390.

WINDISCH, S. (1960) - Die hefeartigen Pilze. Sonderdruck aus: Die Hefen, Bd. 1: 23-208. Nürnberg.

WOLF, E. (1954) - Beitrag zur Systematik der Gattung Mortierella und Mortierella-Arten als Mycorrhizapilze bei Ericaceen. Zentbl. Bakt. Parasitkde, Abt. II, 107: 523-548.

WOLF, F. A. (1929) - The relationship of Microstroma juglandis. J. Elisha Mitchell scient. Soc. 45: 130-135.

WOLF, F. A. (1935) - Morphology of Polythrincium causing sooty blotch of clover. Mycologia 27: 58-73.

WOLLENWEBER, H. W. & H. HOCHAPFEL (1936) - Beiträge zur Kenntnis parasitärer und saprophytischer Pilze. I. Phomopsis, Dendrophoma, Phoma und Ascochyta und ihre Beziehung zur Fruchtfäule. Z. Parasitenk. 8: 561-605.

WOLLENWEBER, H. W. & O. A. REINKING (1935) - Die Fusarien. Berlin, 355 pp.

YAMAMOTO, W. (1959) - Some species of Cladosporium from Japan. Scient. Rep. Hyogo Univ. Agr. 4: 1-6.

YARROW, D. (1969) - Candida steatolytica sp. n. Antonie van Leeuwenhoek 35: 24-28.

YARROW, D. (1970) - Selenotila peltata comb. n. Antonie van Leeuwenhoek 35: 418 (vol. 1969).

YOKOYAMA, T. & K. TUBAKI (1971) - Cultural and taxonomical studies on the genus Actinopelte. Res. Comm. Inst. Ferment. Osaka, 5: 33-77.

ZAMBETAKKIS, C. (1954) - Recherches sur la systématique des Sphaeropsidales - Phaeodidymae. Bull. Soc. mycol. Fr. 70: 219-350.

ZIEGLER, A. W. (1958) - New water molds from Florida. Mycologia 50: 403-406.

ZINDEREN–BAKKER, E. M. van (1934) - Stephanoma tetracoccum spec. nov. Ann. mycol. 32: 101-104.

ZUCK, R. K. (1946) - Isolates between Stachybotrys and Memnoniella. Mycologia 38: 69-76.

ZUNDEL, G. L. (1953) - The Ustilaginales of the world. Penn. St. Univ. Contr. 176, 410 pp.

ZYCHA, H. (1935) - Mucorineae: Krypt.Fl. Mark Brandenbg. 6a. Leipzig, 264 pp.

ZYCHA, H., R. SIEPMANN & G. LINNEMANN (1969) - Mucorales. Lehre, 355 pp.

INDEX TO THE GENERA

Illustrations are marked with an asterisk

Monocephalis 46
Monoceras 173
Monochaetia 166, 175
Monocillium 125, 186, 192, 194*
Monodictys 250, 261*, 262
Monopodium 254
Monopycnis 159
Monospora 64
Monosporella 64
Monostichella 165, 166, 168, 168*
Monotospora 254
Mortierella 36, 48, 49*
Mucor 36, 40*, 41
Mucoricola 45
Muratella 48
Murogenella 249, 259, 261*
Myceliophthora 213, 224
Myceloblastanon 60, 68
Mycelophagus 34
Mycocandida 68
Mycocladus 39
Mycogala 100
Mycogone 127, 247, 251*, 252
Mycokluyveria 68
Mycoleptodiscus 166, 170
Mycorhynchus 107, 123
Mycosphaerella 133, 137, 139*, 150, 152,
    222, 224, 239, 240
Mycosylva 212, 224
Mycotorula 69
Mycotoruloides 69
Mycotypha 38, 49, 49*
Mycovellosiella 224
Myiophyton 50, 50*, 51
Myriococcum 268
Myrioconium 83, 189, 193
Myriosclerotinia 83
Myrothecium 189, 199, 200*
Myxolibertella 158
Myxotrichum 88, 99, 101*

Nadsonia 8, 52, 59*, 61
Nadsioniomyces 64
Naegeliella 29
Naemospora 164, 173
Nakataea 216, 238
Nalanthamala 186, 199
Nannizzia 89, 95, 256
Napicladium 209
Narasimhella 88, 95
Naumoviella 48
Nectaromyces 68
Nectria 108, 121, 122*, 192, 199, 200,
    201, 205, 206
Nectriella 121
Nectriopsis 108, 123, 205

Nematoctonus 177, 178*
Nematogonium 212, 228
Nematomyces 192
Nematospora 53, 63*, 64
Neocosmospora 108, 120*, 121
Neogymnoascus 95
Neofabraea 83
Neomichelia 261
Neorehmia 125
Neosartorya 87, 92, 93*, 94, 196
Neotestudina 132, 137
Neottiospora 160
Neottiosporella 189, 199
Neozimmermannia 123
Nephrospora 115
Neurospora 106, 117*, 119
Niesslia 108, 125
Nigrosabulum 90, 99, 102*
Nigrosphaeria 114
Nigrospora 129, 248, 255, 257*
Nodulosphaeria 133, 143
Nodulisporium 131, 217, 233, 234*
Nomuraea 187, 196
Notarisiella 121
Nowakowskiella 16, 17
Nozemia 34
Nummularia 233

Ocellaria 84, 171
Ochroconis 219, 233, 235*
Octomyxa 15, 15*
Octospora 76, 81
Octosporomyces 58
Oedocephalum 76, 214, 227*, 229
Oedogoniomyces 16, 19
Oidiodendron 98, 180, 181, 182*
Oidium 222
Olpidiella 17
Olpidiopsis 25, 27*, 28
Olpidium 16, 17
Olpitrichum 248
Onygena 90, 95
Oosporidium 66, 69
Oosporoidea 183
Ophiobolus 134, 141
Ophionectria 266
Ophiostoma 56, 104, 111, 229
Oplotheciopsis 125
Oplothecium 125
Orbicula 75, 79
Orbilia 236
Ornithascus 78
Otthia 152
Ovularia 216, 222, 223*

Paecilomyces 91, 92*, 94, 187, 195*, 196

Pachybasidiella 171
Pachybasium 201
Pachysolen 53, 58
Pachyspora 135
Papularia 129, 212, 225*, 246
Papulaspora 268, 269*
Paracoccidioides 256
Paraphaeosphaeria 134, 143, 154
Parasaccharomyces 68
Parasitella 36, 41
Paratorulopsis 66
Parendomyces 68
Passalora 137, 220, 240
Peckiella 109, 127
Pectinotrichum 89, 95
Pedilospora 236
Pellicularia 270
Penicillifer 186, 203*, 205
Penicilliopsis 94, 196
Penicillium 94, 96*, 187, 196
Periconia 212, 225*, 226
Periconiella 220, 240
Pericystis 103
Perisporium 135
Peristomium 115
Peronospora 4
Pesotum 218
Pestalotia 166, 172*, 174
Pestalotiopsis 174
Petalosporus 95
Petasospora 60
Petersenia 25, 28
Petrakia 167, 174*, 175, 249
Petriella 105, 115, 116*
Petriellidium 105, 115
Peyronellaea 149
Peyronellula 99
Pezicula 83, 160, 171
Peziza 75, 81, 228, 229
Pezizula 78
Phacidiopycnis 147, 160
Phacidium 160
Phaeangium 84
Phaeoisaria 218, 232
Phaeoisariopsis 220, 240
Phaeophomopsis 154
Phaeopolynema 161
Phaeoramularia 224
Phaeoscopulariopsis 208
Phaeosphaeria 134, 143
Phaeostoma 106, 114
Phaeotrichum 132, 135, 136*
Phialocephala 188, 202, 207*
Phialomyces 187, 196
Phialophora 8, 85, 118, 126*, 186, 189,
    193, 194*

Phialotubus 187, 197
Philocopra 119
Phloeophthora 34
Phloeosporella 85, 166, 168, 169
Phlyctaena 148, 160
Phlyctochytrium 16, 18, 18*
Pholiota 2, 177
Phoma 137, 146, 149. 150*
Phomatospora 109, ¦ ·
Phomatosporopsis 126
Phomopsella 158
Phomopsis 128, 147, 157, 158*
Phragmodothella 131
Phycomyces 36, 41
Phyllosticta 135, 147, 153*, 155
Phyllostictina 155
Phymatotrichopsis 214, 228
Physalidium 247, 262
Physalospora 108, 110, 129, 246
Physarum 1
Physoderma 16
Phytophthora 4, 25, 33*, 34
Pichia 53, 57*, 60
Pilaira 35, 45
Pilidiella 147, 154
Pilidium 147, 161
Pilobolus 35, 45
Pionnotes 205
Piptocephalis 37, 45
Pirella 35, 41
Pirobasidium 85
Pithoascus 105, 115
Pithomyces 249, 261, 261*
Pityrosporum 65, 67
Placopeziza 84
Plasmodiophora 15
Plectophoma 149
Plectosphaerella 108, 124*, 125, 205
Pleiochaeta 216, 239
Plenodomus 149
Pleocystidium 28
Pleolpidium 19
Pleonectria 122
Pleosphaeropsis 152
Pleospora 134, 142*, 144, 151, 242
Pleotrachelus 16, 18
Pleurage 119
Pleuroascus 90, 100
Pleurophomella 83, 146, 160
Pleurophragmium 233
Pleurosordaria 119
Pochonia 201
Podoconis 259
Podospora 107, 117*, 119
Podostroma 108, 123, 202
Pollaccia 137, 190, 210, 249